T0204153

REVERBERATION CHAMBERS

REVERBERATION CHAMBERS

THEORY AND APPLICATIONS TO EMC AND ANTENNA MEASUREMENTS

Stephen J. Boyes
Defence Science & Technology Laboratory, Fort Halstead, UK

Yi Huang
The University of Liverpool, UK

This edition first published 2016 © 2016 by John Wiley & Sons, Ltd.

Registered Office
John Wiley & Sons, Ltd, The Atrium, Southern Gate, Chichester, West Sussex, PO19 8SQ, United Kingdom

For details of our global editorial offices, for customer services and for information about how to apply for permission to reuse the copyright material in this book please see our website at www.wiley.com.

The right of the authors to be identified as the authors of this work has been asserted in accordance with the Copyright, Designs and Patents Act 1988.

Library of Congress Cataloging-in-Publication Data

Boyes, Stephen J., author.
  Reverberation chambers : theory and applications to EMC and antenna measurements / Stephen J. Boyes, Yi Huang.
    pages cm
  Includes bibliographical references and index.
  ISBN 978-1-118-90624-8 (cloth)
1. Antennas (Electronics)–Design and construction–Technique.   2. Electromagnetic measurements.
3. Radio frequency oscillators.   4. Electromagnetic compatibility.   I. Huang, Yi, (Electrical engineer),
author.   II. Title.
  TK7871.6.B69 2016
  621.382′4–dc23
                                                    2015033560
A catalogue record for this book is available from the British Library.

Set in 10.5/13pt Times by SPi Global, Pondicherry, India
Printed and bound in Singapore by Markono Print Media Pte Ltd

1   2016

Contents

About the Authors

Dr. Stephen J. Boyes received his BEng (Hons) in electronics and communications engineering, MSc (Eng) in microelectronic systems and telecommunications and PhD in antennas/electromagnetics, all from the University of Liverpool, United Kingdom.

His academic research activities and PhD at the University of Liverpool was centred on reverberation chambers, with an emphasis on a wide variety of antennas and antenna measurements for communication applications. His academic research work covered novel textile antennas, multiple input multiple output (MIMO) antennas and novel antenna arrays.

In addition to work in academia, Dr. Boyes has also held various posts throughout 10 years of his working in the industry. In 2013, Dr. Boyes joined the Defence Science & Technology Laboratory where he is now a senior RF Research Scientist leading all antennas and electromagnetics research activities within his group. His present research interests and activities are associated with novel antennas for military applications which include body-worn and vehicle-borne types. He also leads research into novel frequency-selective surfaces and holds numerous patents in both these areas. Dr. Boyes has published many papers in leading international journals and conferences and is also a reviewer for leading academic journals. He also participates and sits on major international committees on behalf of the United Kingdom.

Prof. Yi Huang received his BSc in Physics (Wuhan, China), MSc (Eng) in Microwave Engineering (Nanjing, China) and DPhil in Communications from the University of Oxford, United Kingdom, in 1994. He has been conducting research in the areas of radio communications, applied electromagnetics, radar and antennas since 1986. He has been interested in reverberation chamber theory and applications since 1991. His experience includes 3 years spent with NRIET (China) as a Radar Engineer and various periods with the Universities of Birmingham, Oxford and Essex as a member of research staff. He worked as a Research Fellow at British Telecom Labs in 1994 and then joined the Department of Electrical Engineering & Electronics, University of Liverpool, as a Lecturer in 1995, where he is now the Chair in Wireless Engineering, the Head of High Frequency Engineering Group and the Deputy Head of the Department.

Dr. Huang has published over 250 refereed research papers in leading international journals and conference proceedings and is the principal author of the popular book *Antennas: From Theory to Practice* (John Wiley & Sons, 1st edition, 2008; 2nd edition, 2016). He has received many research grants from research councils, government agencies, charity, EU and industry; acted as a consultant to various companies; served on a number of national and international technical committees (such as the UK Location & Timing KTN, IET, EPSRC and European ACE); and has been an Editor, Associate Editor or Guest Editor of four international journals. He has been a keynote/ invited speaker and organiser of many conferences and workshops (e.g. IEEE iWAT, WiCom and Oxford International Engineering Programmes). He is at present the Editor-in-Chief of *Wireless Engineering and Technology* (ISSN 2152-2294/2152-2308), Leader of Focus Area D of European COST-IC0603 (Antennas and Sensors), Executive Committee Member of the IET Electromagnetics PN, UK national representative to the Management Committee of European COST Action IC1102 (VISTA), a Senior Member of IEEE and a Fellow of IET.

Acknowledgements

The authors would like to thank the following people who have either directly or indirectly helped with the production of this book. Their time and efforts have been greatly appreciated.

Qian Xu – The University of Liverpool, UK

Dr. Y. David Zhang – The University of Manchester, UK

Dr. Ping Jack Soh – K. U. Leuven, Belgium (now Universiti Malaysia Perlis)

Professor Anthony Brown – The University of Manchester, UK

Professor Guy A. E. Vandenbosch – K. U. Leuven, Belgium

Professor David J. Edwards – The University of Oxford, UK

Professor Per-Simon Kildal – Chalmers University of Technology, Sweden

Sandra Grayson – Wiley, UK

Victoria Taylor – Wiley, UK

Teresa Netzler – Wiley, UK

Liz Wingett – Wiley, UK

Sandeep Kumar – SPi Global, India

1

Introduction

1.1 Background

The concept of the Reverberation Chamber (RC) was first proposed by H. A. Mendes in 1968 as a novel means for electromagnetic field strength measurements [1]. The RC can be characterised as an electrically large shielded metallic enclosure with a metallic stirrer to change the field inside the chamber that is designed to work in an 'over-mode' condition (i.e. many modes). It has taken some time for the facility to gain universal acceptance, but by the 1990s, their use for performing Electromagnetic Compatibility (EMC) and Electromagnetic Interference (EMI) measurements was well established and various aspects were studied [2–11]. An international standard on using the RC for conducting EMC testing and measurements was published in 2003 [12]. The RC is now used for radiated emission measurements and radiated immunity tests, as well as for shielding effectiveness measurements. It was in this role that the facility was known for a long time and in part still continues to be [13, 14]. More recently the RC has been employed for antenna measurements due to the rapid development of wireless communications.

It is clear that the role and function of wireless technology in everyday life have reached unprecedented levels as compared to 20–30 years ago. For this change to take place, it has meant that antenna designs and their characterisation have had to evolve also. A question exists as to how the RC has risen to prominence to be proposed

Reverberation Chambers: Theory and Applications to EMC and Antenna Measurements, First Edition.
Stephen J. Boyes and Yi Huang.
© 2016 John Wiley & Sons, Ltd. Published 2016 by John Wiley & Sons, Ltd.

and also to be used for antenna measurements, which represents a brand new capability for the chamber that diverges from its initial intended use. To answer this question, we must partly examine the nature of antenna designs and their operational use.

Traditionally, antennas have always been orientated, and their communication channels configured in a Line of Sight (LoS) manner. For example, we have terrestrial antennas mounted on roof tops, and other directive types of antenna that are employed in satellite communications. The characterisation of these types of antenna for use in LoS communications are widely defined by the application of an equivalent free space reflection-free environment, which is typified by the Anechoic Chamber (AC). In a real application environment, reflection, scattering and diffraction effects may still exist to a certain extent, which brings about the creation of additional wave paths within the communication channel. However, the AC is still the preferred environment to characterise these types of antennas as their radiation patterns (and other subsequent parameters of interest) are of prime importance to the LoS scenario.

When we consider the modern mobile terminals (such as the mobile/cell phone), they do not operate under the premise of an LoS scenario. The antennas inside mobile phones might seldom 'see' the base station and they are expected to work perfectly in Non-Line of Sight (NLoS) environments. This type of environment will readily give rise to signals that will be exposed to reflections caused by large smooth objects, diffraction effects caused by the edges of sharp objects and scattering effects caused by small or irregular objects. When these effects occur, they will cause the creation of additional wave paths which will eventually add at the receiving side. These wave contributions have independent complex amplitudes (i.e. magnitude and phase information), such that at recombination, they may add constructively or destructively or anything in between these extremes. The wave paths and their complex amplitudes are also subject to rapid changes with time, with the terminal moving or parts of the environment (communication channel) changing. This brings about variations in the signal at the receiver and is commonly referred to as fading. The largest variations occur when there is a complete block on the LoS, which is more accurately referred to as small- scale fading, as opposed to large-scale fading, which is usually applied to variations only in the distance from the transmitter or due to part shadowing [15].

As it is important to characterise any antenna in a manner befitting its operational scenario or intended use to accurately reflect the performance merits, another measurement facility is required that can emulate this type of fading environment – this is where the RC comes in. The antenna measurement inside an RC can be closer to a real-world scenario than inside an AC. Furthermore, some measurements, such as the antenna radiation efficiency measurement, may be more efficient and accurate if preformed in an RC than in an AC as we will see later in the book.

When the regulations of the American Federal Communications Commission (FCC) released the unlicensed use of the Ultra Wide-Band (UWB) frequency domain between 3.1 and 10.6 GHz in 2002 [16], a vast amount of interest and industrial/academic research followed. The RC offers little restriction concerning large operational frequencies while also allowing for a vast range of device types and sizes.

Financial implications have also played a part in the subsequent rise to prominence of the RC. With the construction and operation of an AC (here we will only specifically compare with far field ACs), large amounts of anechoic absorber must be purchased to line the walls in order to suppress reflections, and this can be expensive. The RC requires no such absorber, leaving the walls purely metallic to actively encourage reflections – a trait which allows for a cost saving. Furthermore, if one was to compare the relative size of each facility against the lowest frequency of operation, it is possible to conclude that the RC can be constructed smaller in size than its far field anechoic counterpart – again offering a potential cost saving.

The RC is a unique, stand-alone facility, in the sense that it will allow a user full control over the time frame and uncertainty inherent in a given measurement; this distinguishes itself from any other facility. The operational principles for the RC allows for the measurement resolution to be clearly defined which in turn controls the overall measurement time. Mathematical procedures can be defined which link the expected uncertainty to the resolution and thus the time frame. Therefore, before a measurement commences all the parameters can be defined accordingly.

Perhaps one of the more important factors that led to the increasing popularity of the RC concerns the ease of measurement. Due to its unique operation of multiple reflected waves, the angle of arrival of these waves reaching any receiving antenna or device is uniformly distributed over three-dimensional space [17]. What this effectively means is that the angle of arrival and wave polarisation is equally probable which can simplify the characterisation of any device, as in such an environment, their performance is then insensitive to their orientation – this aspect is particularly acute for EMC and antenna measurements.

1.2   This Book

The RC is a very powerful tool for EMC and antenna measurements and has many advantages to offer. There have been many journal articles published over the years on the subject which signifies its increasing popularity; however, few published reference books exist. The most relevant ones are probably references [18] and [19]. They have provided an excellent and comprehensive coverage on the electromagnetic theory on cavities and RCs, and a very good introduction to EMC tests and measurements using an RC. But very limited information is on the RC design and its application to antenna measurements. There are still some important issues on how to use the RC for EMC and antenna measurements in practice.

This book is different from other works on RCs. It is designed to encompass both EMC and antenna measurements together which is important as there are subtleties between how the RC facility is used with respect to these different domains – it is crucial therefore to understand what these are and how to apply the operational principles of the chamber for the benefit of the intended measurement.

The book is also designed to take a reader from the very basic theory of the chamber to a more complicated stirrer design and measurements. It is written to include both detailed theory and practical measurements so a reader can appreciate and understand not only what the theory means, but also how to apply and configure it in a practical manner to complete a desired measurement. With all this information in one place, this book aims to ensure that the reader has a comprehensive yet compact reference source so that the RC can be studied and understood without needing to access a number of different sources which may not be well correlated.

The material covered in this book is underpinned by accepted theory on the RC that is published worldwide along with our own detailed research which covers many of the very latest and cutting-edge trends. The information in the book is also used as part of the Antennas (antenna measurements section) and EMC modules for students at the University of Liverpool. This subsequently is also where all of the measurement work for the book has been wholly conducted. A major feature of this book is to apply the theory to practice.

The book is organised as follows:

Chapter 2: Reverberation Chamber Cavity Theory. This chapter details all the important theoretical concepts that uphold and support the use of the RC as a measurement facility. In this chapter, all the theories are explained, all equations detailed and these are supplemented with practically measured quantities from the RC at the University of Liverpool, to illustrate the magnitude of the quantities derived.

Chapter 3: Mechanical Stirrer Designs and Chamber Performance Evaluation. This chapter presents a general method that can be employed to go through the process of designing mechanical stirring paddles for use in the chamber. New mechanical stirring paddles are designed and presented in this chapter. Also detailed are the complete equations and practical procedures of how the performance of any given chamber may be assessed in an accurate and robust manner.

Chapter 4: EMC Measurements Inside Reverberation Chambers. This chapter focuses specifically on EMC tests in RCs. The relevant standards for EMC tests in RCs are discussed, after which immunity and emission tests are introduced. Practical procedures of how to conduct EMC tests are explained which is followed by practical tests. A comparison between RCs and ACs for EMC radiated emissions is also presented to benchmark both facilities.

Chapter 5: Single Port Antenna Measurements. This chapter is dedicated solely to single port antenna measurements in RCs. The measurements are based around some of the very latest trends in the antenna field with the use of textile antenna which is selected as an example to demonstrate measurement procedures. This chapter not only shows how to measure single port antenna quantities in free space conditions, but also shows how body worn antennas can be measured that include the use of live human beings in the chamber. The radiation efficiency is the major concern of the measurement. This chapter also includes some of the subtle measurement issues that should be avoided when conducting such measurement work in addition to a comprehensive uncertainty assessment, including both procedures and equations.

Chapter 6: Multiport and Array Antennas. This chapter discusses how multiport and array antennas can be measured using the RC. The multiport section includes all measurement procedures and equations for quantities such as diversity gain, correlation and channel capacity and details the performance merits of a new multiport (diversity) antenna for Multiple Input Multiple Output (MIMO) applications. The array section shows how the efficiency of large-scale arrays can be measured using the chamber and develops a new equation to allow this characterisation to take place.

Chapter 7: Further Applications and Developments. This chapter presents and discusses some of the very latest research in RCs that includes how to measure antenna performance parameters without reference antennas. Also included is the use of the RCs for emulating different 'channel' characteristics which is important for many over- the-air measurements.

References

[1] H. A. Mendes, 'A New Approach to Electromagnetic Field-Strength Measurements in Shielded Enclosures', Wescon Tech. Papers, Wescon Electronic Show and Convention, Los Angeles, August 1968.

[2] P. Corona, G. Latmiral, E. Paolini and L. Piccioli, 'Use of reverberating enclosure for measurements of radiated power in the microwave range', *IEEE Transactions on Electromagnetic Compatibility*, vol. 18, pp. 54–59, 1976.

[3] P. F. Wilson and M. T. Ma, 'Techniques for measuring the electromagnetic shielding effectiveness of materials. II. Near-field source simulation', *IEEE Transactions on Electromagnetic Compatibility*, vol. 30, pp. 251–259, 1988.

[4] Y. Huang and D. J. Edwards, 'A novel reverberating chamber: the source-stirred chamber', *Eighth International Conference on Electromagnetic Compatibility, 1992*, IET, 21–24 September 1992, Edinburgh, pp. 120–124.

[5] J. Page, 'Stirred mode reverberation chambers for EMC emission measurements and radio type approvals or organised chaos', *Ninth International Conference on Electromagnetic Compatibility, 1994*, IET, 5–7 September 1994, Manchester, pp. 313–320.

[6] D. A. Hill, 'Spatial correlation function for fields in a reverberation chamber', *IEEE Transactions on Electromagnetic Compatibility*, vol. 37, p. 138, 1995.

[7] T. H. Lehman, G. J. Freyer, M. L. Crawford and M. O. Hatfield, 'Recent developments relevant to implementation of a hybrid TEM cell/reverberation chamber HIRF test facility', *16th Digital Avionics Systems Conference, 1997*, AIAA/IEEE, vol. 1, pp. 4.2-26–4.2-30, 1997.

[8] M. O. Hatfield, M. B. Slocum, E. A. Godfrey and G. J. Freyer, 'Investigations to extend the lower frequency limit of reverberation chambers', *IEEE International Symposium on Electromagnetic Compatibility*, vol. 1, pp. 20–23, 1998.

[9] L. Scott, 'Mode-stir measurement techniques for EMC theory and operation', *IEE Colloquium on Antenna Measurements*, vol. 254, pp. 8/1–8/7, 1998.

[10] G. H. Koepke and J. M. Ladbury, 'New electric field expressions for EMC testing in a reverberation chamber', *Proceedings of the 17th Digital Avionics Systems Conference, 1998*, The AIAA/IEEE/SAE, vol. 1, IEEE, 31 October–7 November 1998, Bellevue, WA, pp. D53/1–D53/6.

[11] Y. Huang, 'Triangular screened chambers for EMC tests', *Measurement Science and Technology*, vol. 10, pp. 121–124, 1999.

[12] IEC 61000-4-21: 'Electromagnetic compatibility (EMC) Part 4-21: Testing and measurement techniques – Reverberation chamber test methods', ed, 2003.

[13] L. R. Arnaut, 'Time-domain measurement and analysis of mechanical step transitions on mode-tuned reverberation chamber: characterisation of instantaneous field', *IEEE Transactions on Electromagnetic Compatibility*, vol. 49, pp. 772–784, 2007.

[14] V. Rajamani, C. F. Bunting and J. C. West, 'Stirred-mode operation of reverberation chamber for EMC testing', *IEEE Transactions on Electromagnetic Compatibility*, vol. 61, pp. 2759–2764, 2012.

[15] P. S. Kildal, *Foundations of Antennas: A Unified Approach*, Lund: Studentlitteratur, 2000.

[16] F. C. Commission, 'Revision of Part 15 of the Commission's Rules regarding Ultra-Wideband Transmission Systems, First Report and Order', ET Docket 98-153, FCC 02-48, 1–118, February 14, 2002.

[17] K. Rosengren and P. S. Kildal, 'Study of distributions of modes and plane waves in reverberation chambers for the characterization of antennas in a multipath environment', *Microwave and Optical Technology Letters*, vol. 30, pp. 386–391, 2001.

[18] D. A. Hill, *Electromagnetic Fields in Cavities*, Hoboken, NJ: John Wiley & Sons, Inc., 2009.

[19] P. Besnier and B. D'Amoulin, *Electromagnetic Reverberation Chambers*, Hoboken, NJ: John Wiley & Sons, Inc., 2011.

2

Reverberation Chamber Cavity Theory

2.1 Introduction

A grounding in the theoretical principles concerning Reverberation Chambers (RCs) is important and essential for any practical undertaking to be completed accurately. The overall field of electromagnetics that contains the foundations of RC cavity theory is a well studied and classical area, and we owe much to authors such as Harrington [1], Balanis [2], Jackson [3], Kraus [4] and more recently Hill [5] for their excellent work in this area. We should also not forget the contribution made by James Clark Maxwell who provided the unified theory and equations that made work in the entire field possible, and Henrich Hertz who practically validated Maxwell's theorems by demonstrating the existence of electromagnetic radiation.

The purpose of this chapter is to present a review and discuss the important theoretical concepts that uphold and support the RCs' use as a measurement facility. We will firstly present a review of the concept of resonant modes and show how these can be used to calculate the electromagnetic fields in an RC. Next, we will show how RCs uniquely deal with the nature of the fields inside the facility and discuss the main principles of operation. The concepts of angle of arrival of plane waves, mode bandwidths and chamber quality (Q) factors will then be introduced and explained, which will be backed up with practically measured data.

Reverberation Chambers: Theory and Applications to EMC and Antenna Measurements, First Edition.
Stephen J. Boyes and Yi Huang.
© 2016 John Wiley & Sons, Ltd. Published 2016 by John Wiley & Sons, Ltd.

A discussion on the statistical forms will be also included in this chapter and a practical way of deducing the statistical forms will be illustrated. At all stages, where appropriate, figures depicting any measurement set-ups and measurement details will be provided to clearly show how all quantities are practically acquired. By the end of this chapter, the reader will have a firm grasp of the important theoretical concepts and be able to appreciate how they impact upon practical measurements.

2.2   Cavity Modes and Electromagnetic Fields

Since most RCs are rectangular in shape, this discussion will be centred solely on rectangular-shaped cavities. It is well known that a metallic rectangular-shaped cavity is resonant when its dimensions satisfy the condition

$$k_{mnp}^2 = \left(\frac{m\pi}{a}\right)^2 + \left(\frac{n\pi}{b}\right)^2 + \left(\frac{p\pi}{d}\right)^2 \tag{2.1}$$

where k_{mnp} is an eigenvalue to be determined; m, n and p are integer numbers and a, b and d are the chamber width, height and length in metres, respectively.

For convenience (2.1) can also be expressed as:

$$k_{mnp}^2 = k_x^2 + k_y^2 + k_z^2 \tag{2.2}$$

where

$$k_x = \left(\frac{m\pi}{a}\right), k_y = \left(\frac{n\pi}{b}\right), k_z = \left(\frac{p\pi}{d}\right) \tag{2.3}$$

It is stated that the simplest method to construct the resonant modes for a rectangular cavity is to derive modes that are Transverse Electric (TE) or Transverse Magnetic (TM) to one of the three axes [5]. The standard convention in this sense is normally chosen to be the z axis with respect to Figure 2.1.

The TE modes are referred to as magnetic modes as the E_z field component is zero. Similarly, the TM modes are referred to as electric modes because the H_z field component is also zero [5]. The corresponding fields in a rectangular cavity can be given as follows, beginning first with the TM components. Please note that the equations presented at this stage do not take any current source into account.

$$E_{zmnp}^{TM} = E_O \sin\left(\frac{m\pi}{a}x\right)\sin\left(\frac{n\pi}{b}y\right)\cos\left(\frac{p\pi}{d}z\right) \tag{2.4}$$

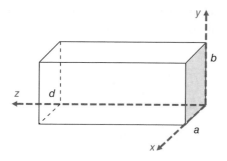

Figure 2.1   Rectangular cavity principle axis.

where E_O is a constant with units of V/m [5]. The transverse components can be issued as follows

$$E_{xmnp}^{TM} = \frac{k_x k_z E_O}{k_{mnp}^2 - k_z^2} \cos\left(\frac{m\pi}{a} x\right) \sin\left(\frac{n\pi}{b} y\right) \sin\left(\frac{p\pi}{d} z\right) \qquad (2.5)$$

$$E_{ymnp}^{TM} = \frac{k_y k_z E_O}{k_{mnp}^2 - k_z^2} \sin\left(\frac{m\pi}{a} x\right) \cos\left(\frac{n\pi}{b} y\right) \sin\left(\frac{p\pi}{d} z\right) \qquad (2.6)$$

The z component of the magnetic field is zero (owing to the definition of the TM mode), and the transverse components of the magnetic field can be given as [5]:

$$H_{xmnp}^{TM} = -\frac{i\omega_{mnp}\varepsilon k_y E_O}{k_{mnp}^2 - k_z^2} \sin\left(\frac{m\pi}{a} x\right) \cos\left(\frac{n\pi}{b} y\right) \cos\left(\frac{p\pi}{d} z\right) \qquad (2.7)$$

$$H_{ymnp}^{TM} = \frac{i\omega_{mnp}\varepsilon k_x E_O}{k_{mnp}^2 - k_z^2} \cos\left(\frac{m\pi}{a} x\right) \sin\left(\frac{n\pi}{b} y\right) \cos\left(\frac{p\pi}{d} z\right) \qquad (2.8)$$

where ε is the permittivity of the media inside the cavity and ω is the angular frequency. By virtue that E_{zmnp}^{TM} be non-zero (Eq. 2.4), the allowable values of the mode coefficients are $m=1,2,3,\ldots; n=1,2,3,\ldots; p=0,1,2,3,\ldots$.

The *TE* (magnetic modes) can be derived in an equivalent manner. The z component of the magnetic field satisfies the scalar Helmholtz equation and the boundary conditions are such that the following is prevalent:

$$H_{zmnp}^{TE} = H_O \cos\left(\frac{m\pi}{a} x\right) \cos\left(\frac{n\pi}{b} y\right) \sin\left(\frac{p\pi}{d} z\right) \qquad (2.9)$$

where H_O is a constant of units A/m. The eigenvalues and axial wave numbers are the same as given in Equations (2.1)–(2.3). The transverse components can now be issued as [5]:

$$H^{TE}_{xmnp} = -\frac{H_o k_x k_y}{k^2_{mnp} - k^2_z} \sin\left(\frac{m\pi}{a}x\right)\cos\left(\frac{n\pi}{b}y\right)\cos\left(\frac{p\pi}{d}z\right) \tag{2.10}$$

$$H^{TE}_{ymnp} = \frac{H_o k_y k_z}{k^2_{mnp} - k^2_z} \cos\left(\frac{m\pi}{a}x\right)\sin\left(\frac{n\pi}{b}y\right)\sin\left(\frac{p\pi}{d}z\right) \tag{2.11}$$

The z component of the electric field is zero by definition of a TE mode and the transverse components of the electric field can be stated as [5]:

$$E^{TE}_{xmnp} = -\frac{i\omega_{mnp}\mu k_y H_O}{k^2_{mnp} - k^2_z} \cos\left(\frac{m\pi}{a}x\right)\sin\left(\frac{n\pi}{b}y\right)\sin\left(\frac{p\pi}{d}z\right) \tag{2.12}$$

$$E^{TE}_{ymnp} = \frac{i\omega_{mnp}\mu k_x H_O}{k^2_{mnp} - k^2_z} \sin\left(\frac{m\pi}{a}x\right)\cos\left(\frac{n\pi}{b}y\right)\sin\left(\frac{p\pi}{d}z\right) \tag{2.13}$$

The allowable mode coefficients in this regard are $m=0,1,2,3,\ldots$; $n=0,1,2,3,\ldots$ and $p=1,2,3,\ldots$; with the only exception that $m=n=0$ is not allowed.

The resonant frequencies f with respect to each individual set of mode coefficients can be deduced by (2.14), adopting the earlier defined notation.

$$f_{mnp} = \frac{1}{2\sqrt{\mu\varepsilon}}\sqrt{\left(\frac{m}{a}\right)^2 + \left(\frac{n}{b}\right)^2 + \left(\frac{p}{d}\right)^2} \tag{2.14}$$

where μ, ε are the permeability and the permittivity of the medium inside the cavity, respectively.

When assessing the number of modes that are present in a given cavity, three common methods exist. The first is based on what is termed 'mode counting', which can be performed by the repeated solution of (2.1) for both TE and TM modes, giving the total number of modes present with eigenvalues less than or equal to k (a practical limit for the propagation of modes). The second method is by an approximation referred to as 'Weyl's formula' [5], which is valid for cavities of general shape and is given in (2.15).

$$N = \frac{8\pi}{3}(a\times b\times d)\frac{f^3}{c^3} \tag{2.15}$$

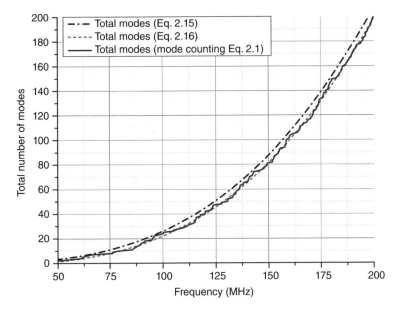

Figure 2.2   Mode numbers vs frequency for the University of Liverpool RC.

where N is the number of modes, f is the frequency in Hertz and c is the wave speed in the cavity medium in meter per second.

The third method is an extension to Weyl's formula specific to rectangular cavities [5] and is stated in (2.16).

$$N = \frac{8\pi}{3}(a \times b \times d)\frac{f^3}{c^3} - (a+b+d) + \frac{1}{2} \qquad (2.16)$$

A comparison of the modal numbers present in the University of Liverpool RC (dimensions of $a=3.6\,\mathrm{m}$, $b=4\,\mathrm{m}$ and $d=5.8\,\mathrm{m}$) by all three methods is shown in Figure 2.2.

From Figure 2.2 we see that the extra terms in (2.16) improve the agreement obtained with the numerical mode counting as opposed to using the original Weyl's formula in (2.15). For increasing frequency, it is seen that the number of modes increases with respect to the cavity volume and the third power of frequency.

The mode density is another important parameter to be able to assess. This quantity charts the amount of modes available in a small bandwidth about a given frequency [5]. This quantity can be found by differentiating (2.16) and is issued in (2.17) after consultation with [5].

$$D_S(f) = 8\pi(a \times b \times d)\frac{f^2}{c^3} - \frac{a+b+d}{c} \qquad (2.17)$$

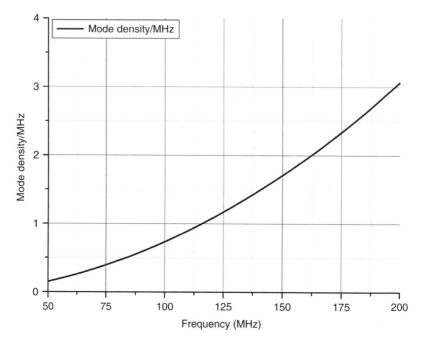

Figure 2.3   Mode density/MHz vs frequency in the University of Liverpool RC.

A plot of the mode density per megahertz vs frequency can be viewed in Figure 2.3. From Figure 2.3 we see that the chamber has a mode density of at least one mode per megahertz from 116 MHz upwards.

A low mode density in any given chamber means that chamber would not have adequate performance, as the mode density is too small to obtain spatial field uniformity [5].

Knowledge of the modal condition inside any given RC is important. As can be seen from Equations (2.4) to (2.13), the fields inside a given cavity can be depicted in terms of resonant modes and their subsequent integer coefficients (m, n and p) satisfying the boundary conditions at the walls of the chamber. In operational practice, different modes are required to be excited in order to promote a sufficient change in the field distribution (to achieve spatial uniformity). The first five allowable modes in the University of Liverpool RC can be seen in Table 2.1.

It is advantageous to be able to calculate and visualise the fields created inside a given chamber by the excitation of resonant modes. However, different from the prior field equations in (2.4)–(2.13), it is advantageous to visualise these fields in a 'non-empty' chamber; that is, with a realistic excitation involved. This provides confidence to be able to interpret the conditions inside the chamber and link them to operational

Table 2.1 The first five resonant modes in University of Liverpool RC.

Distinct mode	m	n	p	Resonant frequency (MHz)
1	0	1	1	45.55
2	1	0	1	49.04
3	1	1	0	56.06
4	1	1	1	61.73
5	0	1	2	63.89

conditions that are witnessed in practice. With respect to the modes and the subsequent fields, it is known from [6] that:

1. The Electromagnetic (EM) fields inside a rectangular chamber outside of the source area are purely the superposition of all TE and TM modes generated within it. In the source area, an extra term needs to be added from the contribution of other hybrid modes.
2. For a current with any polarisation inside the chamber, an electric field with three components may be generated. This essentially means that a signal transmitted by an antenna with one polarisation can be received by an antenna with any polarisation, which is advantageous for measurement purposes.
3. Modes inside the chamber can be controlled by the choice of the polarisation and location of the excitation source which is important as we will see later in this chapter.

In Ref. [6] a computationally efficient series of equations based on cavity Green's functions was derived in order to study the resultant electric fields inside shielded enclosures. The Green's function is essentially a compact means of describing an electric or magnetic field distribution due to a current source, and is desirable here because a realistic source type can be incorporated. The main equations can be stated as (2.18), (2.19), (2.20) and (2.21).

$$E = \frac{1}{j\omega\varepsilon} \int_{source} \underline{G} \cdot J\left(x',y',z'\right) dv' \tag{2.18}$$

where $J(x',y',z')$ = the excitation current, ω is the angular frequency in radians and $\underline{G}$ = dyadic Green's function.

$$E = \frac{1}{j\omega\varepsilon} \int_{source} \left[G_{xy}\hat{x} + G_{yy}\hat{y} + G_{zy}\hat{z} \right] J\left(x',y',z'\right) dv' \tag{2.19}$$

$$\underline{G} = \sum_{p=0}^{\infty} \sum_{m=0}^{\infty} \frac{2\varepsilon_{0m}}{da\alpha \sin \alpha b} \sin k_z z' \cos k_x x'$$

$$\left[\begin{cases} k_z k_x \cos k_z z \sin k_x x \; \begin{array}{l} \sin \alpha y \sin \alpha (b-y') \\ \sin \alpha y' \sin \alpha (b-y) \end{array} \begin{array}{l} y < y' \\ y > y' \end{array} \hat{z}\hat{x} \\ + (k_x^2 - k^2) \sin k_z z \cos k_x x \; \begin{array}{l} \sin \alpha y \sin \alpha (b-y') \\ \sin \alpha y' \sin \alpha (b-y) \end{array} \begin{array}{l} y < y' \\ y > y' \end{array} \hat{x}\hat{x} \\ + k_x \alpha \sin k_z z \sin k_x x \; \begin{array}{l} \cos \alpha y \sin \alpha (b-y') \\ -\sin \alpha y' \cos \alpha (b-y) \end{array} \begin{array}{l} y < y' \\ y > y' \end{array} \hat{y}\hat{x} \end{cases} \right]$$

$$+ \sum_{m=0}^{\infty} \sum_{n=0}^{\infty} \frac{2\varepsilon_{0n}}{ab\beta \sin \beta d} \sin k_x x' \cos k_y y'$$

$$\left[\begin{cases} k_x k_y \cos k_x x \sin k_y y \; \begin{array}{l} \sin \beta z \sin \beta (d-z') \\ \sin \beta z' \sin \beta (d-z) \end{array} \begin{array}{l} z < z' \\ z > z' \end{array} \hat{x}\hat{y} \\ + (k_y^2 - k^2) \sin k_x x \cos k_y y \; \begin{array}{l} \sin \beta z \sin \beta (d-z') \\ \sin \beta z' \sin \beta (d-z) \end{array} \begin{array}{l} z < z' \\ z > z' \end{array} \hat{y}\hat{y} \\ + k_y \beta \sin k_x x \sin k_y y \; \begin{array}{l} \cos \beta z \sin \beta (d-z') \\ -\sin \beta z' \cos \beta (d-z) \end{array} \begin{array}{l} z < z' \\ z > z' \end{array} \hat{z}\hat{y} \end{cases} \right]$$

$$+ \sum_{n=0}^{\infty} \sum_{p=0}^{\infty} \frac{2\varepsilon_{0p}}{bd\gamma \sin \gamma a} \sin k_y y' \cos k_z z'$$

$$\left[\begin{cases} k_y k_z \cos k_y y \sin k_z z \; \begin{array}{l} \sin \gamma x \sin \gamma (a-x') \\ \sin \gamma x' \sin \gamma (a-x) \end{array} \begin{array}{l} x < x' \\ x > x' \end{array} \hat{y}\hat{z} \\ + (k_z^2 - k^2) \sin k_y y \cos k_z z \; \begin{array}{l} \sin \gamma x \sin \gamma (a-x') \\ \sin \gamma x' \sin \gamma (a-x) \end{array} \begin{array}{l} x < x' \\ x > x' \end{array} \hat{z}\hat{z} \\ + k_z \gamma \sin k_y y \sin k_z z \; \begin{array}{l} \cos \gamma x \sin \gamma (a-x') \\ -\sin \gamma x' \cos \gamma (a-x) \end{array} \begin{array}{l} x < x' \\ x > x' \end{array} \hat{x}\hat{z} \end{cases} \right] \qquad (2.20)$$

where $\varepsilon_{0n} = \begin{array}{l} 1 \text{ when } n = 0, \\ 2 \text{ when } n \neq 0, \end{array}$ $\alpha = \sqrt{k^2 - k_z^2 - k_x^2}$, $\beta = \sqrt{k^2 - k_x^2 - k_y^2}$ and $\gamma = \sqrt{k^2 - k_y^2 - k_z^2}$

Now, assuming a y polarised unit current (with respect to the principle axis defined in Figure 2.1, meaning that the current source is vertically linearly polarised), and with reference to (2.18), (2.19) and (2.20), the resultant E_y electric field on an xz plane can be obtained using (2.21).

$$E = \frac{1}{j\omega\varepsilon}\sum_{m=0}^{\infty}\sum_{n=0}^{\infty}\frac{2\varepsilon_{0n}}{ab\beta\sin\beta d}\sin k_x x' \cos k_y y'$$

$$\begin{bmatrix} \left\{ k_x k_y \cos k_x x \sin k_y y \begin{array}{l} \sin\beta z \sin\beta(d-z') \rbrack z < z' \\ \sin\beta z' \sin\beta(d-z) \rbrack z > z' \end{array} \hat{x} \right\} \\ +\left\{ \left(k_y^2 - k^2\right)\sin k_x x \cos k_y y \begin{array}{l} \sin\beta z \sin\beta(d-z') \rbrack z < z' \\ \sin\beta z' \sin\beta(d-z) \rbrack z > z' \end{array} \hat{y} \right\} \\ +\left\{ k_y\beta \sin k_x x \sin k_y y \begin{array}{l} \cos\beta z \sin\beta(d-z') \rbrack z < z' \\ -\sin\beta z' \cos\beta(d-z) \rbrack z > z' \end{array} \hat{z} \right\} \end{bmatrix} \qquad (2.21)$$

The following series of figures chart the E_y electric field distribution on an xz plane in the University of Liverpool RC as a function of varying frequency. The observation location (y) in all the depicted plots is selected as $b/2$; that is, the mid-point in height of the chamber.

The coordinate locations for the source in all plots are $x1=0.4$ m, $x2=0.5$ m, $y1=1.35$ m, $y2=1.4$ m and $z=0.5$ m, which situates the source towards one corner of the chamber in the principle x axis, and locates the source between 1.35 and 1.4 m high from the chamber floor. These selections have been made because this is the typical source location chosen in the chamber during practical measurements, as will be seen in later chapters.

From Figures 2.4, 2.5, 2.6, 2.7 and 2.8 we can see that the electric fields inside the shielded chamber are formed as a result of standing waves that have a sine and cosine dependence. Also, it is evident that as the frequency increases, the fields begin to vary

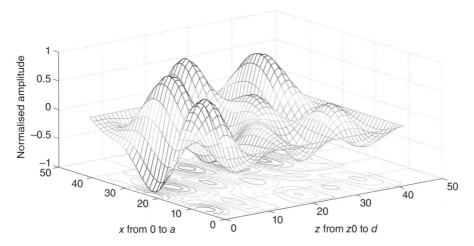

Figure 2.4   Normalised E_y field distribution in University of Liverpool RC at 200 MHz.

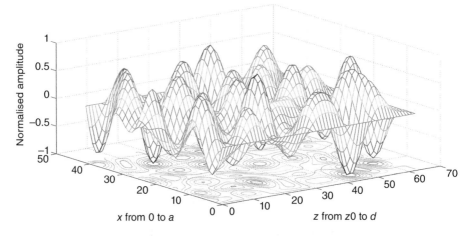

Figure 2.5 Normalised E_y field distribution in University of Liverpool RC at 400 MHz.

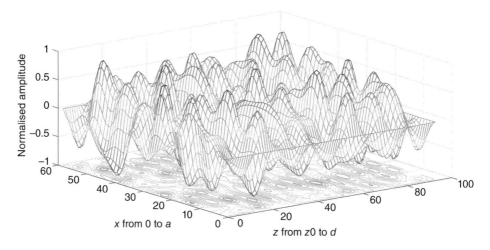

Figure 2.6 Normalised E_y field distribution in University of Liverpool RC at 600 MHz.

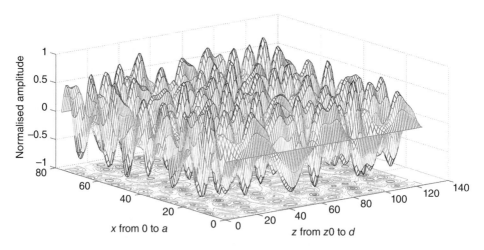

Figure 2.7 Normalised E_y field distribution in University of Liverpool RC at 800 MHz.

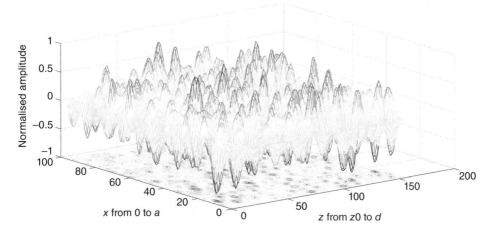

Figure 2.8 Normalised E_y field distribution in University of Liverpool RC at 1000 MHz.

in a more complex manner. At all frequency values plotted here, the electric field magnitude is seen to have significantly different values.

This is not the desired property for operation inside the chamber. For example, if an antenna was to be placed in such a field distribution during a given experiment, then the amount of power received would be dependent upon the location in which the antenna was situated. Any results taken therefore would not necessarily be repeatable and/or correct. In practice, a uniform field distribution is desired in order to provide a consistent environment in which to conduct measurements.

2.3 Mode Stirring Techniques

Having established the nature of the electric field inside a shielded enclosure resulting from a representative current source, and expressed the desire and need for a uniform field distribution, we will now discuss the techniques available to achieve the desired uniform field. One of the unique features of the RC is that it employs 'Mode Stirring techniques' or simply 'Stirring techniques' to deal with the nature of the field as previously shown.

The purpose of these techniques is to be able to render the field distribution statistically uniform and isotropic on average; in achieving this, we could potentially situate an antenna or any piece of test equipment in different locations inside the chamber (potentially even in different orientations) and be able to receive a similar consistent average power level (providing that all other measurement variables remained as constant). This would at least go some way towards providing a statistically repeatable facility. The question now is, what are these techniques and how are they performed?

2.3.1 Mechanical Stirring

This mode stirring technique is perhaps considered to be the most common technique employed to permute the fields inside an RC. The action of mechanical stirring is performed by the rotation or movement of electrically large paddles or plates situated inside the confines of the chamber, whose shape or design is usually configured to be non-symmetrically shaped.

The paddles/plates are usually configured to be rotated/moved in a step wise or continuous manner and upon each rotation or movement; the boundary conditions for the electromagnetic field inside the chamber are changed [7]. The effect of changing the boundary conditions will mean that the 'peaks' and 'nulls' evident from Figures 2.4, 2.5, 2.6, 2.7 and 2.8 will change as a function of 'observation' location.

This will essentially mean that the fields can be rendered statistically homogeneous and isotropic on average from many stirrer increments, provided that the field location is situated approximately $\lambda/2$ (where λ is the free space wavelength in metres) from the chamber boundaries (e.g. the walls) [8]. The rotation or movement of the plates/paddles can also give rise to statistical environments which can be advantageous to measurements in the chamber. More on the statistical aspects can be found later in this chapter.

For now, an example of the mechanical stirring paddle design in the University of Liverpool RC can be viewed in Figure 2.9. It can be seen that two principle sets of

Horizontal stirrers

Vertical stirrers

Figure 2.9 Example of mechanical stirrer design.

mechanical stirring paddles exist; one set mounted from a central rotational shaft from the floor to the roof, the other set mounted from the front to the back wall at ceiling height.

The requirement for two separate sets of mechanical stirring paddles arises from any potential change in polarisation from reflected waves inside the chamber, regardless of source polarisation. This way, both polarisation characteristics emitting from the source (assuming linear polarisation) should be effectively stirred; helping to ensure that the polarisation characteristic of any reflected wave reaching an antenna or test object or in the reciprocal case of emissions from a test object to receiving device, has, on average, a desired equal probability and can thus be considered 'un-polarised' (no dominance one way or another).

The paddles are typically sized and are designed to operate in the frequency domain where they can be considered 'electrically large', that is, they are at least comparable to the wavelength of operation and the cavity can generally be considered to be over-moded (i.e. many available modes). When this is not the case the paddles' performance in significantly changing the field distribution will diminish, thus representing one limitation in the process.

2.3.2   Polarisation Stirring

Despite the adoption of two principle sets of mechanical stirring paddles and the theoretical validation concerning the un-polarised nature of the RC, a bias can poten-tially exist in practice. It was reported in Refs [9, 10] that a differing amount of TE and TM modes could potentially be excited during measurements, which subse-quently manifests itself as a difference in received power with respect to different receiver antenna polarisations. In Ref. [9], when discussed purely from a wave and modal perspective, investigations were present that cited differences from 3 to 9 dB could be evident.

This mode stirring mechanism advocates that both vertical and horizontal transmit polarisations (assuming linear polarisations) be measured and an average taken from both measured results to remove any effect of bias, regardless of any receiver polarisa-tion. That is, it aims to ensure that no 'polarisation imbalance' exists as it can represent a serious source of measurement uncertainty for antenna measurements in the RC.

In addition, a polarisation imbalance could also be prevalent as a result of different performances in the vertical and horizontal mechanical stirrer paddles; hence, they have a different efficiency in permuting the fields in the chamber (stirrer efficiency). This necessitates that the design and measurement validation of mechanical stirrer paddles should be carefully and rigorously conducted as we will see in Chapter 3.

This stirring mechanism and this type of bias could perhaps serve to highlight the differing requirements between antenna measurements and Electromagnetic Compatibility (EMC)-type measurements that are conducted using the RC. For EMC, the equipment under test may not have a preferred or designed orientation with respect to the polarisation of any radiated emissions or for any incoming waves when immunity testing is being conducted. For the measurement of antenna quantities, the antennas involved usually have a predesigned and preferred polarisation and the bias could be more sensitive here with respect to any results obtained.

In practice, a large number of samples are also normally required to obtain accurate average values. It stands to reason that the more samples one has to form an average, the more accurate that average value will be. Therefore, this technique is also useful in generating an additional number of measurement samples and should be used in conjunction with the mechanical stirring aspect previously defined.

In conclusion, a systematic polarisation imbalance in practice could be caused by the excitation of different amounts of TE and TM modes, in addition to having mechanical stirrer paddles with different stirrer efficiencies. The technique of measuring two transmit polarisations regardless of the receiver orientation can remove the problem [9, 10]. The technique can also be a useful method in generating an additional number of measurement samples, normally required for accuracy purposes/lower the measurement uncertainty.

2.3.3 Platform and Position Stirring

Platform stirring advocates the movement of an Antenna under Test (AUT) or Device under Test (DUT) to different locations in the chamber by means of a turntable, and an overall average is taken from measured samples at each measurement location. This technique was devised in Ref. [11] as a means of improving accuracy in a small RC, to avoid progressive samples from different locations being correlated (thus sufficiently independent), it necessitates that movement distances to each different receiver locations should be no less than $\lambda/2$. Position stirring is essentially the same procedure. However, this technique does not employ the use of a turntable, instead it relies on the physical movement of the AUT/DUT to different locations manually. The advantage of these procedures is that it has been shown to improve accuracy in measurements by giving rise to an additional number of independent measurements for use in the averaging process.

2.3.4 Frequency Stirring

Frequency or electronic stirring advocates the taking of a further average from acquired measurements in an RC, by the use of an algorithm in the post- processing stage. The frequency stirring application has been discussed at length (albeit for a 2D cavity) in Ref. [5] as a means of improving the spatial uniformity of fields and decreasing the interaction between a line source and the chamber walls.

The application advocates that the smoothing bandwidth (*defined as the total number of points in a given window from which to take a further successive average*) should decrease with an increase in frequency (increase in modal density) in a given chamber.

Hence, a higher level of frequency stirring is required at lower modal densities because it is more difficult to achieve field uniformity in a sparse modal environment for reasons previously discussed.

A degree of care is normally required when performing frequency stirring because taking too large an average (large smoothing bandwidth) can result in the loss of frequency resolution in any measured data. For example, when measuring the radiation efficiencies or complex reflection coefficients of antennas in an RC, the amount of frequency stirring is usually chosen to be much less than the absolute operating bandwidths of the antenna in order to avoid this problem. This technique is typically always used in order to obtain smooth and precise results, particularly so for antenna measurements.

2.4   Plane Wave Angle of Arrival

The purpose of this subsection is to theoretically examine the Angle of Arrival (AoA) nature of plane waves inside the University of Liverpool RC. This section is important to be able to verify because if we could confirm the characteristics of the propagation scenario, it would go some way to demonstrate the suitability of the RC as a multipath emulator for mobile communications and also for antenna measurements in general. It is also useful to theoretically demonstrate that a multiple reflected wave environment can allow for a simplification in any antenna characterisation, as the performance is insensitive to the orientation. For EMC-type measurements, this type of result is useful in providing the theoretical proof that a device under test in immunity testing, for example, is being illuminated in all directions and angles to thoroughly test its operational performance.

Now, with respect to earlier Equations (2.1), (2.2), (2.4) and (2.9), it has been shown in Ref. [12] that we can represent the sine and cosine terms as exponentials via Euler's relationship and express the product in the form of (2.22).

$$
\begin{aligned}
&\cos(u) \cdot \cos(v) \cdot \sin(w) \\
&= A_{mnp}^{TE} \cdot \frac{e^{ju} + e^{-ju}}{2} \cdot \frac{e^{jv} + e^{-jv}}{2} \cdot \frac{e^{jw} - e^{-jw}}{2i} \\
&= \text{const} \cdot \left\{ \begin{matrix} e^{ju+jv+jw} + e^{ju-jv+jw} + e^{-ju+jv+jw} + e^{-ju-jv+jw} \\ -e^{-ju+jv-jw} - e^{ju-jv-jw} - e^{-ju+jv-jw} - e^{-ju-jv-jw} \end{matrix} \right\} \\
&= \text{const} \cdot \sum e^{\pm ju \pm jv \pm jw}
\end{aligned}
\tag{2.22}
$$

In turn, the specific forms of E_z in (2.4) and H_z in (2.9) can be represented as (2.23).

$$
\left\{ \begin{matrix} H_z^{TE} = \text{const} \cdot \sum e^{-jk\hat{k} \cdot r} \\ E_z^{TM} = \text{const} \cdot \sum e^{-jk\hat{k} \cdot r} \end{matrix} \right\}
\tag{2.23}
$$

where

$$r = x\hat{x} + y\hat{y} + z\hat{z},\ \hat{k} = \left\{ \frac{k_x\hat{x} + k_y\hat{y} + k_z\hat{z}}{k} \right\} \text{ and } k_x = \pm\frac{m\pi}{a},\ k_y = \pm\frac{n\pi}{b},\ k_z = \pm\frac{p\pi}{d}$$

The crucial terms above are the $\pm$ signs that need to be permuted in order to obtain different plane wave terms [12]. Each term above is thus seen to represent an individual plane wave propagating in the $\hat{k}$ direction, meaning that (2.23) symbolises a sum of eight plane waves for both TE and TM modes. The only exception here is when one of the indices is zero; only four plane waves are then evident. Once the allowable and existing modes have been calculated using (2.14) for both TE and TM modes, the corresponding AoA of the plane waves can be calculated from (2.24).

$$\varphi = \arctan\left(\frac{k_y}{k_x}\right),\ \theta = \arctan\left(\frac{\sqrt{k_x^{\,2} + k_y^{\,2}}}{k_z}\right) \tag{2.24}$$

Figure 2.10 depicts a series of figures concerning the AoA of plane waves in the University of Liverpool RC as a function of frequency. The lines displayed on each three-dimensional traces represent the incoming arrival towards an origin (in this case a fictitious unit sphere), and the small patches can be thought of as parts of the associated wavefronts [12] – a simplification in order to visualise.

From Figure 2.10 it can be seen that the AoA comes from every conceivable direction over the unit sphere. With increasing frequency it can also been seen that a higher density of waves are apparent in a smaller band each time. The exact statistical nature of this type of result has been previously studied in Ref. [12], who confirmed that a uniform distribution of AoAs exists, provided that enough modes are excited (sufficiently large band).

A uniform distribution will essentially mean that a plane wave (with subsequent polarisation) can arrive from any direction/angle over the unit sphere with an equal probability. This is an important result to confirm as it proves one of mechanisms as to how measurement procedures in the RC can be simplified. This is because, with such a property, it should not matter too much how a given device or antenna is orientated and the direction and characteristics of any associated radiation patterns. This characteristic should be confirmed and upheld in practice provided that an equal amount of TE and TM modes are excited as discussed earlier, and that no significant and direct line of sight (LoS) path will exist from source to receiver – a topic that will be covered in detail later in this chapter.

Table 2.2 illustrates the number of modes and subsequent number of plane waves in the University of Liverpool RC and confirms that for increasing frequency, we have an ever richer number of modes and plane waves available.

Specifically for antenna measurements in real environments, it should be stressed they will subtly differ from the isotropic nature of the RC with respect to the AoA of waves as proved here. An antenna operating in a real environment may have a preferred

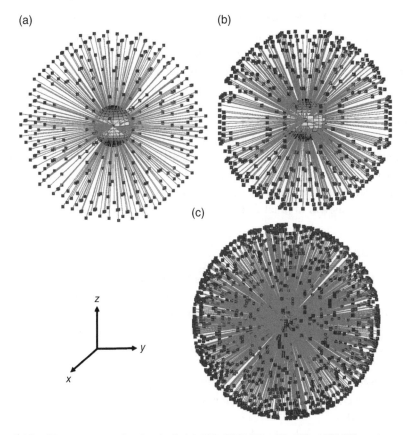

Figure 2.10 Plane wave angle of arrival. (a) 200–225 MHz, (b) 400–410 MHz and (c) 900–905 MHz.

Table 2.2 The number of modes and plane waves in University of Liverpool RC.

Frequency sub-bands (MHz)	Number of modes	Number of plane waves
200–225	87	620
400–410	122	924
900–910	643	4 992
1940–1950	1866	14 776
2400–2410	1930	15 384

orientation, and in outdoor environments they may have a larger probability of waves coming close to horizontal [13].

For mobile telecommunications, real environments can have a larger amount of vertically polarised waves as most base stations are vertically polarised. This can present a cross-polarised power discrimination (XPD), which can differ in different real environments [13]. For measurements in real environments, this could become problematic as any results obtained would be a function of the environment they were measured in and may not always be repeatable.

One of the specific advantages of the RC is that statistically repeatable results have been demonstrated [14], outweighing the fact that the RC emulated isotropic environment has no true counterpart in reality. However, that being said, it can be argued that if a mobile terminal is used in a real environment with multiple arbitrary orientations, the average of those results would approach what is witnessed in the RC [13]; hence, the statistics of any user are taken into account that may include different antenna positions, orientations and varying polarisations.

With respect to the number of modes and plane waves that will be excited during any measurements, this will in practice depend on the value of the 'mode bandwidth'.

2.5 Average Mode Bandwidths

In the RC, theoretically a mode will only be excited when a source excites the chamber at a frequency that corresponds to the exact frequency of the resonant mode [12]. This would imply however that the Quality (Q) factor of the mode is in fact infinite [12]. In reality, the mode Q factor is finite in value and the mode will exhibit a certain 'bandwidth'; thus implying that the modes do not have an infinite cut-off and this gives rise to the term 'bandwidth'.

This 'bandwidth' can be defined as: *'The bandwidth over which the excited power in a particular cavity mode with resonant frequency f_o is larger than half the excited power at f_o.'* [10].

The average mode bandwidth (Δf) can simply be defined as *the average of the mode bandwidths of all of the modes excited about an excitation frequency (f)*, meaning that at a given (f), modes may excited in the range [12]:

$$f - \Delta f / 2 \le f_o \le f + \Delta f / 2 \tag{2.25}$$

where $\Delta f = f / Q$ and Q=the chamber Q factor. The theory and definitions here are analogous to the theory and definitions that can be found in Ref. [15]. Therefore, the average mode bandwidth can be found from knowledge of the measured chamber Q factor or it can be deduced from (2.26) and (2.27) after consultation with [10].

$$\Delta f = \frac{c^3 \times \eta_{\text{TOTAL\_TX}} \times \eta_{\text{TOTAL\_RX}}}{16\pi^2 V f^2 G_{\text{net}}} \tag{2.26}$$

where V=the inside volume of the chamber, $\eta_{\text{TOTAL\_TX}}$ is the total radiation efficiency of the transmitting antenna (*defined as the ratio of total radiated power to the power incident on the antenna port*) and $\eta_{\text{TOTAL\_RX}}$ is the total radiation efficiency of the receiving antenna. G_{net} can be found from (2.27).

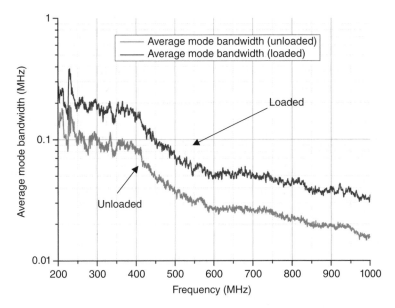

Figure 2.11   Measured average mode bandwidths in University of Liverpool RC.

$$G_{net} = \frac{\left\langle \left|S_{21}\right|^2 \right\rangle}{\left(1-\left|S_{11}\right|^2\right) \times \left(1-\left|S_{22}\right|^2\right)} \qquad (2.27)$$

where $\langle\ \rangle$ symbolises the average of the scattering parameters, S_{21} is the transmission coefficient, and the two terms in the denominator represent the mismatch efficiencies of the transmitting and receiving antennas, respectively. The denominator terms of (2.27) are not signified as being from an ensemble average because the data used here was acquired using an anechoic chamber (AC).

Figure 2.11 depicts both the measured average mode bandwidth in an 'empty' scenario (just containing transmit and receive antennas) and a 'loaded' scenario in the University of Liverpool RC (dimensions $a = 3.6$ m, $b = 4$ m and $d = 5.8$ m).

In both scenarios the receiving antenna is a log-periodic antenna and the transmitting antenna is a large Vivaldi antenna; both with known efficiency values. In the 'loaded' configuration, the chamber has been loaded with two pieces of anechoic absorber in two corners of the chamber to illustrate the effect this has on the average mode bandwidth. The average mode bandwidth is important to know because in measurement situations it governs the channel characteristics in the chamber. For example, it can control whether the fading channel is frequency flat or frequency selective. A further discussion on the channel characteristics with a view to measurements can be found later in this chapter and also in Ref. [16]. The average mode bandwidth and thus the channel characteristics can be controlled by 'loading' the chamber as the results of Figure 2.11 suggest.

From Figure 2.11 it can be seen that the average mode bandwidth is slightly decreasing in size for increasing frequency, albeit in a relatively small scale. Figure 2.11 would suggest that the window in which subsequent modes can be excited grows smaller at higher frequencies. This should not be too problematic however, as the small mode bandwidth will be offset by the high mode density that exists at higher frequencies, meaning that many modes can still be excited. The values of the average mode bandwidth are smaller here as compared with [17] which is a consequence in part of the difference between very large and small RCs. The curves do not appear to be entirely smooth because they have been deduced using measured data with only a finite number of points (718 measured samples).

Furthermore, the decreasing trend of the plot would also suggest that the Q factor of the chamber would increase for increasing frequency from the relationship stated under (2.25). We will move on to assess this aspect (Q factor) next.

2.6   Chamber Quality (Q) Factor

It is well known that the Q factor describes the rate at which a system at resonance loses energy due to its conversion to other forms (e.g. heat). As previously defined in Section 2.5, when considering a single resonance, a low Q system will have a large bandwidth and will dissipate energy very rapidly. When we think of an RC, it is a multimode facility, but the concept of the Q factor is still important as it describes information as to the overall losses that exist within the facility for a given set up.

The initial reference works for the chamber Q factor was published by Corona in 1980 [18] and by Hill in 1994 [19]. Hill picked up from Corona's earlier work and proceeded to derive in detail the multiple mechanisms that contribute to RCs' total Q factor. Hill defined five separate loss terms to exist:

1. Wall losses ($Q1$)
2. Losses due to aperture leakage ($Q2$)
3. Losses due to absorption in any loading objects in the chamber ($Q3$)
4. Antenna losses ($Q4$)
5. Losses due to water vapour absorption ($Q5$ but only at frequencies >18 GHz).

In terms of the mathematics, the total Q factor and the separate loss mechanisms can be briefly summarised as follows [19].

$$Q = \frac{\omega U_s}{P_d} \tag{2.28}$$

where U_s = steady state energy and P_d = power dissipated.

$$U_S = WV \tag{2.29}$$

where $V=$ volume of the cavity and $W=$ energy density.

$$W = \varepsilon_o E^2 \tag{2.30}$$

where $\varepsilon_o=$ permittivity of the medium in the cavity (free space) and $E=$ Root Mean Square (RMS) electric field.

The power density (S_c) in the cavity can also be expressed as (2.31).

$$S_c = \frac{E^2}{\eta_O} \tag{2.31}$$

where $\eta_O=$ intrinsic impedance of the medium in the cavity.

The separate loss terms ($Q1$) to ($Q4$) from [19] can be stated as (2.32), (2.33), (2.34) and (2.35), respectively.

$$Q_1 = \frac{3V}{2\mu_r S\delta} \tag{2.32}$$

where $\delta = \sqrt{\dfrac{1}{\pi f \mu_w \sigma_w}}, \mu_r = \dfrac{\mu_w}{\mu_o}$. μ_w and σ_w are the permeability and conductivity of the wall material, respectively, δ is the skin depth and S is the cavity surface area.

$$Q_2 = \frac{2\pi V}{\lambda \langle \sigma_a \rangle} \tag{2.33}$$

where $\langle \sigma_a \rangle=$ averaged absorption cross section of the medium loading the chamber. For irregular-shaped objects, this term could be complex to solve analytically. A measurement procedure to deduce this term was derived by Carlberg et al. and can be found in Ref. [20].

$$Q_3 = \frac{4\pi V}{\lambda \langle \sigma_1 \rangle} \tag{2.34}$$

where $\langle \sigma_1 \rangle=$ average transmission cross section of any apertures.

$$Q_4 = \frac{16\pi^2 V}{m\lambda^3} \tag{2.35}$$

where $m=$ the impedance mismatch factor or mismatch efficiency ($m=1$ for a matched load).

Under steady state conditions, the power transmitted into the chamber (P_t) is required to be equal to the power dissipated by the loss mechanisms that may exist in a given chamber configuration. Therefore,

$$P_t = P_d \tag{2.36}$$

By substituting (2.28), (2.29), (2.31) into (2.36), the power density in the cavity can be rewritten as (2.37).

$$S_c = \frac{\lambda Q P_t}{2\pi V} \tag{2.37}$$

By use of an impedance-matched antenna, the received power (P_r) is stated in Ref. [19] to be a product of the effective area $\lambda^2/8\pi$ and the received power can be expressed as (2.38).

$$P_r = \frac{\lambda^3 Q}{16\pi^2 V} P_t \tag{2.38}$$

Solving (2.38) in terms of Q yields the chamber Q factor in terms of the measured power ratio (P_r/P_t). Thus (2.39) represents a commonly used measured approach to obtain the chamber Q factor [19]. The equation does assume that highly efficient and well impedance-matched antennas are employed.

$$Q = \frac{16\pi^2 V}{\lambda^3} \times \frac{P_r}{P_t} \tag{2.39}$$

where

$$\frac{P_r}{P_t} = \frac{\left\langle |S_{21}|^2 \right\rangle}{\left(1 - (|S_{11}|)^2\right)\left(1 - (|S_{22}|)^2\right)} \tag{2.40}$$

The terms $(|S_{11}|)$ and $(|S_{22}|)$ represent the reflection coefficients of the transmitting and receiving antennas, respectively. In (2.40) they are assumed to be acquired in an AC, hence the omission of an averaging term. Furthermore, (2.40) also assumes that any cable losses are negligible as a result of calibrating the system.

An alternative representation of the loss terms in (2.32)–(2.35), derived in terms of the average mode bandwidth instead of Q factor of the modes can be found in Ref. [21]. As stated earlier, this is also a representative means of establishing the Q factor of a given chamber. Figure 2.12 depicts the measured Q factor of the University

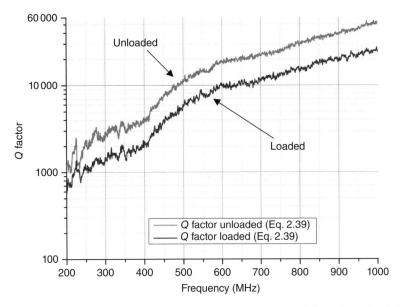

Figure 2.12   Measured Q factor in University of Liverpool RC from Equation 2.39.

of Liverpool RC in both unloaded and loaded scenarios. The results have been acquired from measured data with a finite number of measured samples (718); hence they are not entirely smooth.

A result derived from the average mode bandwidth method is also provided for comparison in Figure 2.13.

Comparing Figures 2.12 and 2.13 we can see that both methods yield similar values for the chamber Q factor. Further, from this similarity, it is possible to conclude that the average mode bandwidth values from Figure 2.11 would appear accurate. It is also worthy to acknowledge at this stage that the measured Q factor values will have some contribution from the antennas used in the measurement (irrespective of efficiency).

A communication in Ref. [7] has discussed this point and concluded that an assessment of the chamber Q factor deduced from the chamber time constant instead of the direct frequency domain method of Equation (2.39) could diminish the contribution of the antennas involved, leaving just the effect of the chamber alone. For completeness, the chamber time constant (τ) is related to the chamber Q factor by the following:

$$\tau = \frac{Q}{2\pi f} \tag{2.41}$$

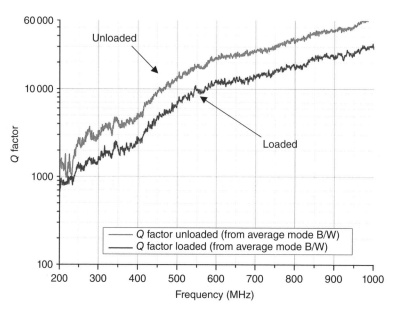

Figure 2.13   Measured Q factor in University of Liverpool RC from average mode bandwidths.

2.7   Statistical Forms

The statistical forms of measured RC data are very important. For example, if we are to consider the modern mobile terminal antenna, the operation of these devices is such that they will seldom 'see' a base station. Therefore, the cellular communication link has to be maintained in a Non-Line of Sight (NLoS) propagation environment and the devices are expected to function flawlessly. The NLoS propagation environment in mobile communications is well known to exhibit Rayleigh fading properties [22], and this would be advantageous to be able to replicate. Successful replication of any realistic operational environment is important for any measurement task, as it is correct that any device should be characterised in a manner that is befitting of its true operation.

The RC is one such statistical emulator and has propagation characteristics similar to that found in the urban and indoor environments. This is how the facility is appropriate for use for measurements requiring a realistic propagation environment.

The statistical forms of RC data have received much treatment over the years. Kostas and Boverie examined the statistical forms in 1991 [23] and concluded that each of the three measured components of the electric field (voltage) manifest themselves as being Rayleigh distributed (which is the same as the chi-squared distribution with 6 degrees of freedom). From there, they concluded that the power density was exponentially distributed [23].

Table 2.3 Statistical forms measurement details.

Parameter	Description
Source antenna	Vivaldi
Receive antenna	Log-periodic (HL223)
Frequencies (MHz)	100–1000
Number of frequency data points	801
Stirring sequences	1 degree mechanical stirring
	Polarisation stirring
Vector Network Analyser (VNA)	Anritsu 37369A
Source power (dBm)	−7
Chamber loading	None

From Ref. [13], it is further stated that as long as the Line of Sight (LoS) is absent and enough plane waves exist, the in-phase and quadrature components of the received complex signal are normally distributed (complex gaussian), and the phase is uniformly distributed over 2π. However, it should be noted that if any significant LoS path were to exist, then the statistical distributions, specifically the magnitudes, will be altered which can have profound consequences. This aspect will be discussed separately in Section 2.8. Equally, if enough plane waves do not exist (cavity does not support enough resonant modes), then the statistical distributions will not necessarily correspond to what is stated here.

For now, the purpose of this subsection is to briefly discuss and examine the theoretical statistical forms in the University of Liverpool RC, to visually assess how the chamber complies with the ideal NLoS theoretical forms discussed earlier. This subsection is further divided into six separate sections to discuss the method of analysis and to individually represent the magnitude, complex signal, power and phase information (respectively) in an NLoS scenario. Concluding remarks and recommendations on the method of analysis are presented in Section 2.7.6.

The statistical forms issued herewith have all been configured with an NLoS path between source and receiver, based upon practical measurements in the University of Liverpool RC. The measurement details for these tests are presented in Table 2.3.

2.7.1 Statistical Methods of Analysis

Two primary methods that can be chosen to visually represent the statistical forms, they are the cumulative distribution function (CDF) and the probability density function (PDF). These two aspects can be theoretically and mathematically defined as follows.

Since a probability cannot be negative, all PDFs must be positive or zero [5].

$$f(g) \geq 0 \tag{2.42}$$

The PDFs are not always continuous or even finite; however, since the random variable g must lie between $-\infty$ and ∞, the relationship in (2.43) must hold [5].

$$\int_{-\infty}^{\infty} f(g) dg = 1 \tag{2.43}$$

From the above definitions, we can assess the CDF properties as the probability P that a random variable G lies between a and b as an integral over f in the form of (2.44).

$$P(a < G \leq b) = \int_{a}^{b} f(g) dg \tag{2.44}$$

Thus, a CDF must exhibit the following:

1. The CDF is a non-decreasing function of g
2. CDF$(-\infty) = 0$
3. CDF$(\infty) = 1$

2.7.2   Statistical Forms of Measured Magnitudes

As previously stated, the ideal form of any magnitudes in an NLoS environment manifests themselves as being statistically Rayleigh distributed. The following plots visually depict the nature of the CDF and the PDF forms of practically measured data against the ideal theoretical distribution. For brevity, only data relating to 600 and 800 MHz is shown here. This selection has been made arbitrarily and has been chosen at frequencies where many modes will exist – hence the cavity can be considered electrically large.

The theoretical Rayleigh distributions are defined in Equations 2.45 and 2.46, according to Ref. [24].

For the CDF,

$$F(x) = \left(1 - e^{\left(-x^2/2\sigma^2\right)}\right), \ x \in [0,\infty] \tag{2.45}$$

and for the PDF,

$$f(x) = \frac{x}{\sigma^2} e^{-x^2/2\sigma^2}, x \geq 0 \tag{2.46}$$

where σ = distribution shape parameter.

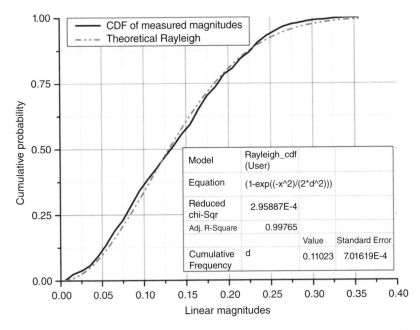

Figure 2.14   CDF of measured magnitudes vs theoretical Rayleigh distribution at 600 MHz.

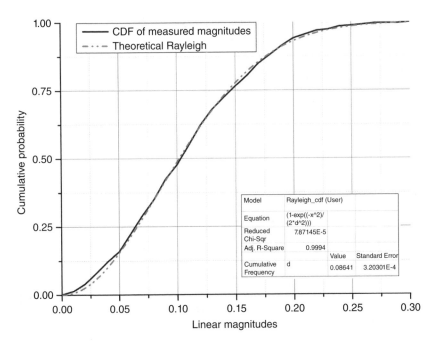

Figure 2.15   CDF of measured magnitudes vs theoretical Rayleigh distribution at 800 MHz.

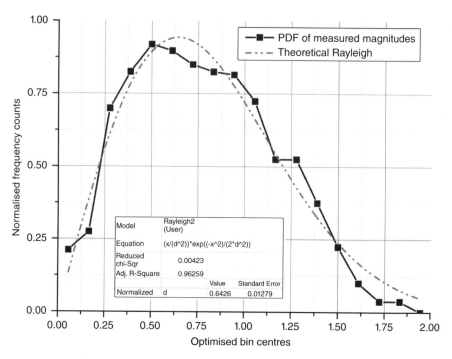

Figure 2.16 PDF of measured magnitudes vs theoretical Rayleigh distribution at 600 MHz.

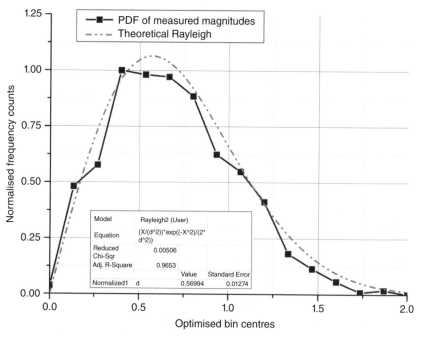

Figure 2.17 PDF of measured magnitudes vs theoretical Rayleigh distribution at 800 MHz.

Please note that at this stage, the purpose of the statistical plots is simply to depict what the distributions look like; hence all plots are presented as a visual means only. We do not intend to imply any significance of the quality of the statistical fit for reasons discussed in Section 2.7.6.

Figures 2.14 and 2.15 portray the CDFs of the measured magnitudes against a theoretical Rayleigh distribution at 600 and 800 MHz, respectively.

The PDFs can be viewed in Figures 2.16 and 2.17 at 600 and 800 MHz, respectively.

2.7.3   Statistical Distribution of Complex Samples

As previously stated, the ideal forms of any complex measured data (magnitude and phase), assuming NLoS and an abundance of modes in the cavity, should manifest themselves as being normally distributed (complex Gaussian). The following plots depict the CDF and PDF forms against their ideal forms as a means of visualisation; again, no significance of the quality of the statistical fit is to be implied. The theoretical normal distributions have been defined in (2.47) and (2.48) according to Ref. [24].

For the CDF,

$$F(x) = \frac{1}{2}\left[1 + \text{erf}\left(\frac{x-\mu}{\sigma\sqrt{2}}\right)\right], x \in \Re \qquad (2.47)$$

where $\text{erf}(x) = \dfrac{2}{\sqrt{\pi}}\displaystyle\int_0^x e^{-t^2}\,dt$ and is defined as the error function, μ = mean and σ^2 in this (normal) distribution will represent the variance.

For the PDF,

$$f(x) = \frac{1}{\sigma\sqrt{2\pi}}e^{-(x-\mu)^2/2\sigma^2}, x \in \Re \qquad (2.48)$$

where μ = mean and σ^2 in this (normal) distribution will represent the variance.

Figures 2.18 and 2.19 depict the CDFs of the measured complex data at 600 and 800 MHz, respectively against their theoretical forms.

The PDF forms of the measured complex samples at 600 and 800 MHz can be seen in Figures 2.20 and 2.21, respectively.

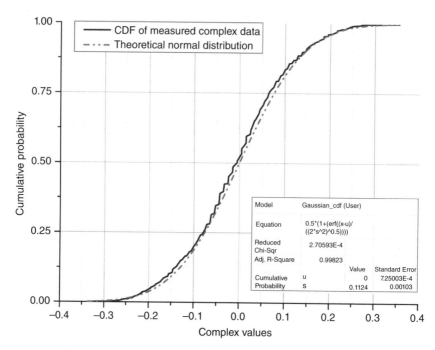

Figure 2.18   CDF of measured complex samples vs theoretical normal distribution at 600 MHz.

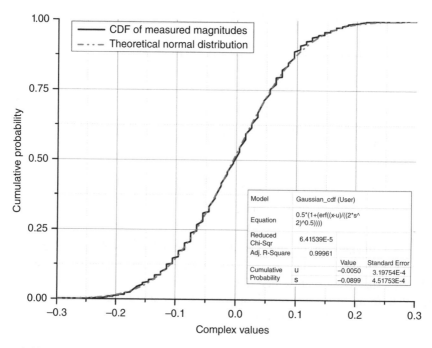

Figure 2.19   CDF of measured complex samples vs theoretical normal distribution at 800 MHz.

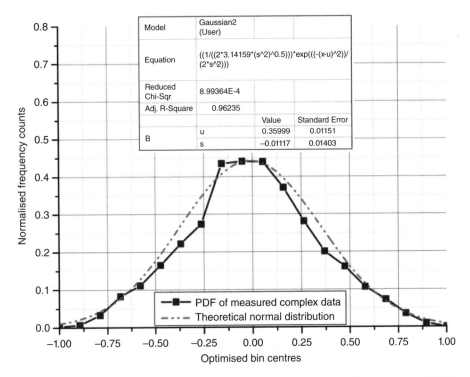

Figure 2.20 PDF of measured complex samples vs theoretical normal distribution at 600 MHz.

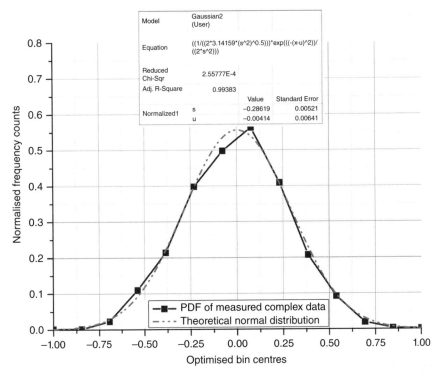

Figure 2.21 PDF of measured complex samples vs theoretical normal distribution at 800 MHz.

2.7.4   Statistical Distribution of Measured Power

Again, as stated previously, the ideal statistical form of any power samples obtained in an NLoS scenario should ideally be exponentially distributed. This subsection discloses the measured obtained power samples against the ideal theoretical CDF and PDF forms. These ideal forms have been defined according to Ref. [24].
For the CDF,

$$F\left(x;\lambda\right)=\begin{cases}1-e^{-\lambda x}, & x\geq0\\ 0, & x<0\end{cases}$$ (2.49)

where λ = distribution rate parameter.
For the PDF,

$$f\left(x;\lambda\right)=\begin{cases}\lambda e^{-\lambda x}, & x\geq0\\ 0, & x<0\end{cases}$$ (2.50)

Figures 2.22 and 2.23 depict the CDFs of the measured power at 600 and 800 MHz, respectively against their theoretical forms.
The PDF forms of the measured power at 600 and 800 MHz can be seen in Figures 2.24 and 2.25, respectively.

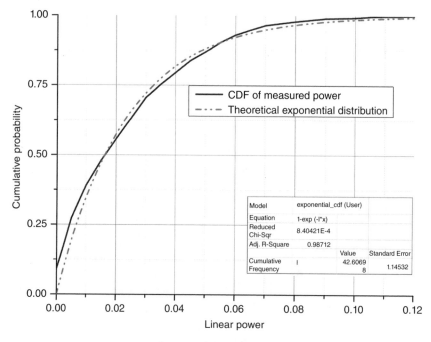

Figure 2.22   CDF of measured power vs theoretical exponential distribution at 600 MHz.

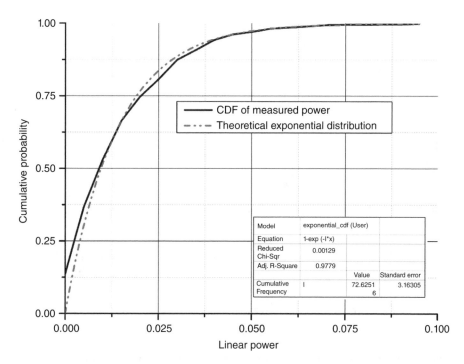

Figure 2.23   CDF of measured power vs theoretical exponential distribution at 800 MHz.

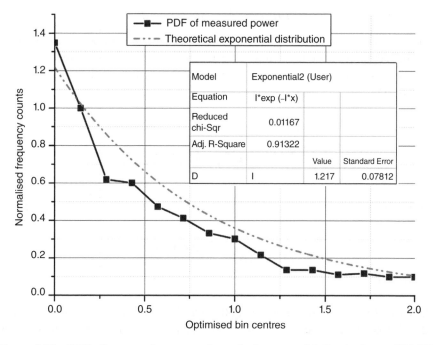

Figure 2.24   PDF of measured power vs theoretical exponential distribution at 600 MHz.

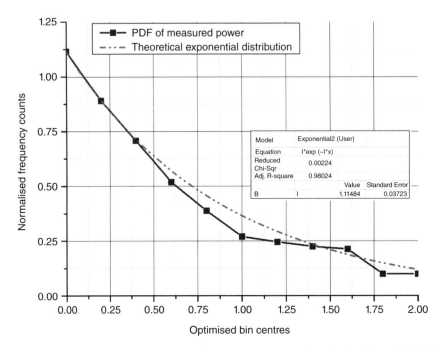

Figure 2.25 PDF of measured power vs theoretical exponential distribution at 800 MHz.

2.7.5 Statistical Distribution of Measured Phase

The phase information obtained from an NLoS scenario should be uniformly distributed. This section tests the uniformity of the measured phase information against their ideal theoretical forms. The theoretical CDF and PDF have been defined as follows [24].
For the CDF,

$$F(x) = \begin{cases} 0, & x \le a \\ \dfrac{x-a}{b-a}, & a < x \le b \\ 1, & x \ge b \end{cases} \tag{2.51}$$

For the PDF,

$$f(x) = \begin{cases} \dfrac{1}{b-a}, & a < x < b \\ 0, & \text{otherwise} \end{cases} \tag{2.52}$$

The CDFs for the measured phase at 600 and 800 MHz can be seen in Figures 2.26 and 2.27, respectively, whereas the PDF forms can be seen in Figures 2.28 and 2.29.

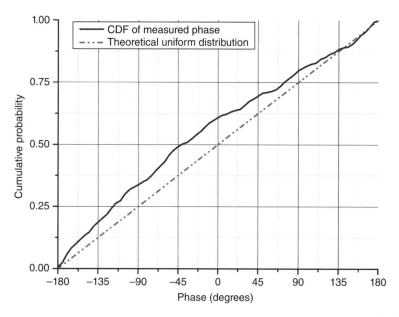

Figure 2.26   CDF of measured phase vs theoretical uniform distribution at 600 MHz.

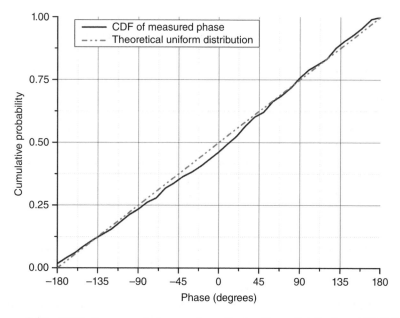

Figure 2.27   CDF of measured phase vs theoretical uniform distribution at 800 MHz.

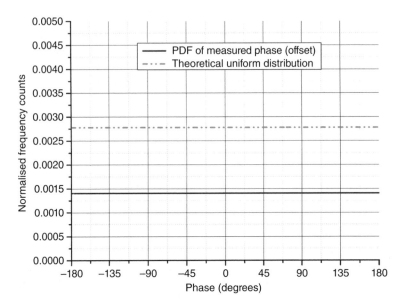

Figure 2.28   PDF of measured phase vs theoretical uniform distribution at 600 MHz.

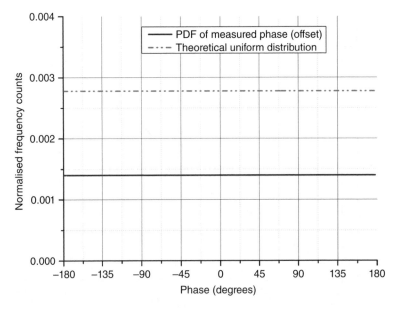

Figure 2.29   PDF of measured phase vs theoretical uniform distribution at 800 MHz.

Please note that in the two PDF figures (Figures 2.28 and 2.29), the measured distribution has been slightly offset from the theoretical distribution in order to visualise both forms clearly.

2.7.6   Concluding Remarks and Recommendations

Assessing the measured and theoretical forms in Figures 2.14–2.29, one could be tempted to conclude that the trends between the measured and theoretical forms appear to be very close indeed, and be thus convinced that the measured data does indeed correspond to the ideal theoretical forms.

However, on the use of such graphical evidence alone, one must attach little weight to the evidence in deriving complete conclusions. Placing sole reliance on coefficient of determination analysis through graphical methods alone could be considered questionable. The reasons for this statement is because the graphical evidence does not take into account the following recommended points.

1. The definition and testing of a null hypothesis
2. The assignment of a confidence interval in testing the null hypothesis
3. The acceptance and rejection of the null hypothesis based on (1) and (2)
4. The probability of a given result upon accepting or rejecting the null hypothesis
5. A comparison to other methods/results

In our case, points (1–5) must also be performed as a function of frequency, which as a consequence, does render the graphical techniques extremely time consuming. Statistical methods do exist, however that attempt to satisfy the constraints highlighted above are aimed at deriving more confident outcomes.

Presenting a thorough discussion on statistical analysis and specific methods goes beyond the aims of this book; however, a brief summary of two existing techniques that aim to satisfy the constraints mentioned above and derive more confident outcomes can be listed as follows.

One such technique is the 'Kolmogorov Smirnov' (KS) test [25]. This test, in a one sample format, is only valid for continuous CDFs and seeks to test the distribution of unknown samples against a hypothesised continuous distribution. When performed with tools such as Matlab™, it will allow for the specification of a null hypothesis and the assignment of confidence intervals, it will also provide results concerning the acceptance or rejection of a null hypothesis and the probability in accepting or rejecting that null hypothesis. The downside to the method is that the hypothesised distribution must be fully specified, which is not always convenient to do, given that any standard deviations, mean values and any other associated parameter concerned with the distribution are liable to change with changing frequency.

One technique that seeks to relax the requirements of the Kolmogorov-Smirnov test was developed by Lilliefors in 1967 and 1969 [26, 27]. The Lilliefors test compares the same test statistic as the Kolmogorov-Smirnov test, but relaxes the constraints on completely specifying the hypothesised continuous distribution. When employed with Matlab™, this statistical test allows for the testing of both normal and exponential distributions and could provide more of a reliable benchmark of statistical significance than the one sample KS test and curve-fitting methods alone; such a procedure and methodology would be recommended. When it comes to the measured magnitudes in an RC, such a statistical test is not necessarily needed; there is a simpler and more straightforward procedure that can be applied: The Rician K factor.

2.8   Line of Sight Elements

Statistically, if any LoS path exists in an RC, the Rayleigh distribution will be altered. The elements of multipath (stirred power) and any LoS path (unstirred power) together will manifest itself as being Rician distributed when the unstirred power becomes more dominant.

A definition of unstirred and stirred power can be stated as [28]:

Unstirred power is the power that is coupled directly from a transmitting antenna to an antenna under test that undergoes minimal reflection or interaction with any mode stirrer.

Stirred power results from power radiating from a transmitting antenna that fully interacts with a mode stirrer. This also includes significant reflections from the chamber walls that interact with the paddle.

A thorough discussion was presented in Ref. [28] which showed how the RC can be used to emulate a Rician environment if desired. This could prove advantageous when characterising the performance of antennas in such environments; for example, when antennas are operating near the mobile base stations.

If Rician emulated environments are desired, the basic premise of operation would be to point the transmitting and receiving antennas at one another as opposed to the standard operation in RCs where the transmitting antenna is purposely directed away from a receiver by either pointing it at the mechanical stirring paddles, directed towards a corner of the chamber or it is hidden behind a blocking plate or shield to prevent any direct path from occurring.

When the transmitting and receiving antennas are pointed directly at one another, the following techniques can be used to configure varying levels of Rician statistical environments, with the possibility of K factors of 10 or greater being experimentally confirmed by these techniques in Ref. [28].

1. Varying the transmit and receive antennas separation from one another
2. Varying the relative azimuth angles of the transmit or receive antenna
3. Varying the polarisation of the transmit or receive antenna (i.e. co-polarised, 45°, cross-polarised)
4. Loading the chamber with varying amounts of anechoic absorber or, for example, a dummy phantom or cylinder(s)

However, for the majority of antenna terminal and EMC measurements in an RC, the minimisation of any unstirred power and the promotion of a Rayleigh environment is of crucial importance for reasons previously discussed. Further, for measured antenna quantities such as radiation and total radiation efficiency, it has been shown that any unstirred power can represent a serious source of measurement uncertainty [29]. For these types of measurements therefore, unstirred elements must be crucially minimised.

The format of any unstirred multipath propagation in the RC has become better known as a Rician K Factor. A full proof relating to the derivation of this quantity can be found in Ref. [28] but is summarised as follows.

Any transmission measurement in an RC essentially initiates a measured transfer function in the chamber comprising of the chamber's physical and statistical properties. This measured transfer function ($|S_{21}|$) or ($|S_{12}|$), assuming the use of scattering parameters (S parameters) and depending on the measurement set up, can be stated to represent the sum of two components: a direct component (d) and a component which constitutes the stirred energy in a chamber (s) [28].

$$\left(|S_{21}| = S_{21d} + S_{21s}\right) \tag{2.53}$$

where S_{21d} = the direct component and S_{21s} = the stirred component. If no stirred component exists (i.e. pure LoS environment) then only a direct component will exist, such as what is present in an anechoic environment.

As the statistical investigation on the complex data yielded; under ideal RC conditions (pure NLoS scenario), the complex transmission measured data should be normally distributed with a zero mean and identical variance.

$$\langle S_{21s} \rangle = 0 \tag{2.54}$$

where $\langle\ \rangle$ signifies the average of the said scattering parameters, and:

$$\mathrm{var}\left[\mathrm{Re}\left(S_{21s}\right)\right] = \mathrm{var}\left[\mathrm{Im}\left(S_{21s}\right)\right] = \left\langle\left[\mathrm{Re}\left(S_{21s}\right)\right]^2\right\rangle = \left\langle\left[\mathrm{Im}\left(S_{21s}\right)\right]^2\right\rangle = \sigma_R^2 \tag{2.55}$$

where σ_R is essentially the standard deviation. The direct component (S_{21d}) on the other hand has a zero variance and a non-zero mean:

$$\mathrm{var}\left[\mathrm{Re}\left(S_{21d}\right)\right]=\mathrm{var}\left[\mathrm{Im}\left(S_{21d}\right)\right]=\left\langle\left[\mathrm{Re}\left(S_{21d}\right)\right]^2\right\rangle=\left\langle\left[\mathrm{Im}\left(S_{21d}\right)\right]^2\right\rangle=0 \quad (2.56)$$

From (2.55) and (2.56), the variance of the real and imaginary components of S_{21} can be given as [28]:

$$\mathrm{var}\left[\mathrm{Re}\left(S_{21}\right)\right]=\mathrm{var}\left[\mathrm{Im}\left(S_{21}\right)\right]=\left\langle\left[\mathrm{Re}\left(S_{21}\right)\right]^2\right\rangle=\left\langle\left[\mathrm{Im}\left(S_{21}\right)\right]^2\right\rangle=\sigma_R^2 \quad (2.57)$$

Alternatively,

$$2\sigma_R^2 = \left\langle\left|S_{21}-\left\langle S_{21}\right\rangle\right|^2\right\rangle \quad (2.58)$$

The mean value of S_{21} is related to the direct component by:

$$d_R = \left|\left\langle S_{21}\right\rangle\right| \quad (2.59)$$

Thus the Rician K factor (ratio of unstirred to stirred power) can be expressed fully as [28]:

$$K = \frac{d_R^2}{2\sigma_R^2} = \frac{\left(\left|\left\langle S_{21}\right\rangle\right|\right)^2}{\left\langle\left|S_{21}-\left\langle S_{21}\right\rangle\right|^2\right\rangle} \quad (2.60)$$

The following figures depict results for the Rician K factor as a function of frequency in the University of Liverpool RC, the quantity being calculated from (2.60).

The measurement results detail frequencies up to and including 1000 MHz, with Table 2.3 again illustrating the measurement parameters used in the investigation. The measurement set-up is discussed in Figure 2.30. The results presented in this subsection are representative of what can be termed the 'average Rican K factor', which is an average of the K factor in (2.60), measured for two transmit polarisations (used in polarisation stirring). The results presented have also been frequency stirred by 25 MHz to improve the readability.

From Figure 2.30, the picture depicts the measurement set up configured in an NLoS scenario, with the horizontally polarised directional transmitting antenna being pointed directly at the mechanical stirring paddles. The directional horizontally polarised log-periodic receiving antenna is also seen to be directed away from the transmitting antenna. This set up aims to minimise the proportion of power that can conceivably

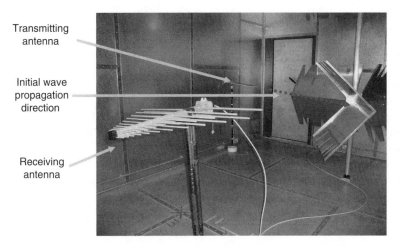

Figure 2.30 Measurement set up for Rician *K* factor tests.

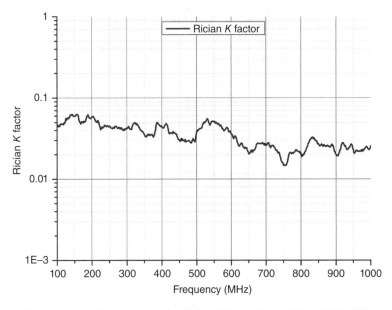

Figure 2.31 Non-line of sight Rician *K* factor from 100 to 1000 MHz.

travel direct from transmitter to receiver with the purpose of the test used to determine how much 'unstirred power' exists under 'standard' operating conditions.

From Figure 2.31 it can be seen that the values obtained for the Rician *K* factor are significantly less than 1 indicating that the stirred power in the chamber is dominant. Figure 2.31 proves that the LoS path is very small and as a result the measured

Table 2.4   Measurement parameters for K factor parallel investigation.

Parameter	Description
Frequencies (MHz)	1000–6000
Frequency data points	1601
Transmitting antenna	Small (homemade) Vivaldi
Receiving antenna (directional)	Rohde and Schwarz dual ridge horn (HF 906)
Receiving antenna (omnidirectional)	Small (homemade) UWB Monopole
Chamber loading	None
Source power (dBm)	−7
Stirring sequences	1 degree mechanical stirring
	Polarisation stirring
	25 MHz frequency stirring

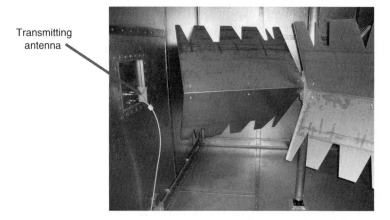

Transmitting antenna

Figure 2.32   Rician K factor measurement set-up 1000 to 6000 MHz.

magnitudes will tend towards a Rayleigh distribution as opposed to being Rician distributed (ideal Rayleigh distribution: $K \to 0$).

Furthermore, the scattered nature of the power will help to yield the ideal forms of the plane wave AoA as discussed earlier in Section 2.4. The result also determines that the placement of the transmitting and receiving antenna does not yield any significant proportion of direct power being received. Nevertheless, a different placement of transmitting antenna warrants inspection as well as the use of omnidirectional antennas on the receiving side to chart the K factor performance.

Table 2.4 details the measurement parameters for these experiments.

Figure 2.32 illustrates the different placement of the transmitting antenna in this (parallel) investigation and Figure 2.33 depicts the overall measurement set-up.

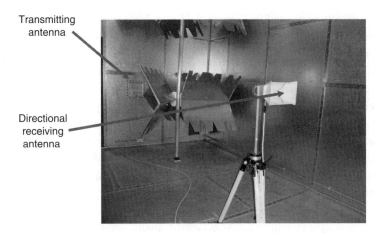

Figure 2.33   Rician *K* factor measurement set-up 1000 to 6000 MHz.

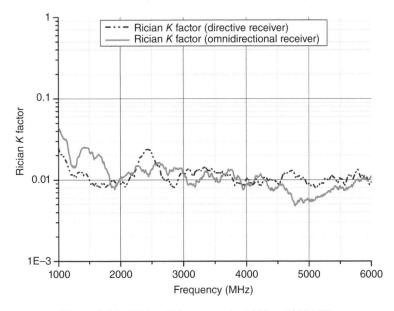

Figure 2.34   Rician *K* factor results 1000 to 6000 MHz.

The small Vivaldi (transmitting) antenna, seen in Figure 2.32, is to be located on a fixed rigid cable right behind the mechanical stirring paddles. The question now is: Will this result in a significantly different *K* factor?

The *K* factor results for the directive receiver (shown in Figure 2.33) and an omnidirectional receiver are presented together in Figure 2.34.

From the results presented in Figures 2.31 and 2.34, it can be seen that neither the different placement of transmitting antenna nor the use of an omnidirectional receiver has had a significant effect on the proportion of direct power; the chamber performs consistently well in this regard over a wide frequency range. Both results provide increased confidence to assume that the chamber corresponds to a Rayleigh environment under 'standard' operating conditions (i.e. with no purposely configured LoS).

We can further test the Rayleigh hypothesis with the assessment of the following scatter plots. The scatter plot is essentially a mathematical diagram using two cartesian coordinates to display given parameters. In this case it is used to gauge the mean distribution of the complex data from the above directional and omnidirectional antenna experiments from 1000 to 6000 MHz.

From Equation 2.54, we have stated and shown that the ideal forms of measured transmission (complex) data in an NLoS environment should normally be distributed with a zero mean and identical variance. Testing the mean of this measured data via the scatter plot should yield a series of samples that are strongly grouped around the centre and as such prove that the stirred power is dominant. Any significant deviations from this would provide evidence to indicate that the unstirred power is dominant and that the magnitudes can therefore deviate from being Rayleigh distributed.

Figures 2.31, 2.34, 2.35 and 2.36 further validate that the chamber emulates the desired NLoS characteristics and shows that the Rician K factor method is a useful and direct way of demonstrating the measured channel statistics in an RC. A few outliers are visible from both scatter plots but these do not appear to be great in number not in magnitude from the ideal zero mean value that is sought.

One point to note at this stage is that the measured channel samples have all been demonstrated in an 'empty' chamber; that is, one that only contains the transmitting and receiving antennas themselves. When characterising the performance of an RC, it is important to take into account the 'loaded' performance also, as during given measurement tasks the chamber may have varying support structures or other items contained inside.

Depending on the amount of loading and the strategic placement of any loading items, this can serve to dampen resonant modes in the chamber. The net result of such an action is that if this loading is severe enough, many resonant modes will be damped which will in turn impact the nature of the field distribution. Such scenarios can mean that as the mode stirring processes are undertaken, the field distribution is not significantly changed as different modes will struggle to be excited. When this is the case, any measurement uncertainties are likely to be high, and the ideal isotropic nature of the plane wave AoA will not exist, meaning that more power can go direct to a given receiver (increase in K factor).

Therefore, we can say that as the RC is loaded in higher and higher quantities, the chamber will go from being a multi-reflective scattered environment to one that begins to approach an anechoic (chamber) type environment. Although significant chamber

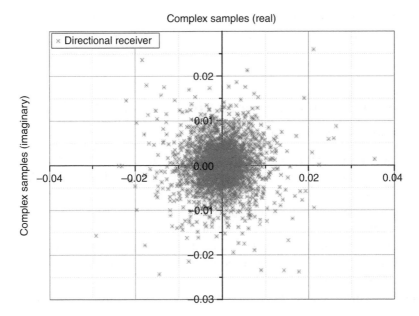

Figure 2.35   Scatter plot for directional receiver from 1000 to 6000 MHz.

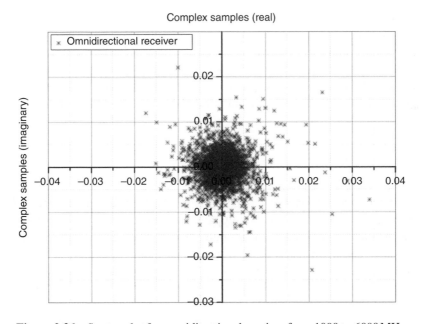

Figure 2.36   Scatter plot for omnidirectional receiver from 1000 to 6000 MHz.

loading may not always be preferred for some antenna measurements, varying the loading of an RC can be very beneficial when channel sounding activities are considered as we will see next.

2.9 Reverberation Chamber as a Radio Propagation Channel

2.9.1 Channel Parameters

Recent research work has been performed to look in greater depth at channel sounding measurements within an RC as a means of further demonstrating the suitability of the facility for wireless Over-the-Air (OTA) measurements [29, 30]. Such work is important as it relates not only how channel parameters can be controlled in the RC, but also how these parameters compare to real world environments. Relating channel parameters back to real world environments is an important step because it can enable more and more device testing to be conducted in the facility, as opposed to conducting all measurements outdoor.

Some of the main parameters of interest for channel sounding activities can be listed as follows.

1. Delay spread
2. Coherence bandwidth
3. Doppler spread
4. Coherence time

The remainder of this chapter will present a brief discussion on these channel sounding parameters and some derivations from literature on how they may be measured and calculated in the RC will also be provided. Where appropriate, measurement results will be presented to illustrate the magnitudes of the quantities taken in the University of Liverpool RC.

2.9.2 Coherence Bandwidth

The coherence bandwidth can be defined as *the frequency range over which the channel is correlated* [30]. It is stated that in multipath environments, it is easier to measure the Root Mean Square (RMS) delay spreads rather than the coherence bandwidth (B_C) [30]. The two quantities are inversely proportional to one another according to Ref. [30] and Equation 2.61.

$$B_C = \frac{1}{k(\sigma_T)} \tag{2.61}$$

where k is a constant that depends on the environment and σ_T is the delay spread. For completeness, the RMS delay spread can be deduced from Equation 2.62 [30].

$$\sigma_T = \sqrt{\frac{\sum_k P(\tau_k)\tau_k^2}{\sum_k P(\tau_k)} - \left(\frac{\sum_k P(\tau_k)\tau_k}{\sum_k P(\tau_k)}\right)^2} \tag{2.62}$$

where $P(\tau) = |h(\tau)|^2$ and the received power $P(\tau_k)$ at the time delay τ_k is the power delay spread and $h(\tau)$ is the impulse response obtained from the inverse Fourier transform (IFT) of the channel frequency response [30].

A thorough discussion was presented in Ref. [30] that experimentally showed that the coherence bandwidth was proportional to the average mode bandwidth (discussed earlier in this chapter), and was equal to it if the coherence bandwidth was properly defined. The term k in Equation 2.61 is also dependent upon how the coherence bandwidth is defined.

The definition in Ref. [30] for the coherence bandwidth states that the bandwidth can be defined in terms of the half bandwidth or the full bandwidth (which is twice as large), at which the complex correlation function has a value of 0.5 or the envelope correlation function has a value of 0.5. In Ref. [30] the definition of the complex correlation function was applied.

From Ref. [30], it is stated that in RCs, $k=2\pi$ when the complex correlation of 0.5 half bandwidth definition is applied or $k=2\sqrt{3}\pi$ when the envelope correlation definition is applied. This was shown to agree well with the measurement evidence that was presented which confirmed that the average mode bandwidths were in fact equal to the coherence bandwidths.

2.9.3   Doppler Shift Frequency

A detailed discussion is presented in Ref. [29] regarding the physical deduction of the Doppler spread that can be emulated in RCs. The Doppler spread can be defined as *the range of Doppler frequencies over which the Doppler spectrum is above a certain threshold* [31]. This is an important quantity to be able to calculate and control since the total isotropic sensitivity of active wireless devices and stations will be affected by the Doppler spread of the propagation channel.

Conventionally, the deduction of the Doppler spread can be found from a direct evaluation defining the Doppler spread. However, it was shown in Ref. [31] that via the use of standard vector network analysers, the Doppler spread can be easily measured and calculated in RCs, eliminating the need to resort to sophisticated signal processing techniques.

This subsection will document the necessary equations for the Doppler spread evaluation and will present some results of the Doppler spreads measured in the University of Liverpool RC.

The method present in Ref. [31] showed that the Doppler frequencies can be deduced using both step wise (individual stirrer increments) and continuous stirring methods. In each case, the S_{21} value is required to be measured according to a given number of stirrer steps or a given stirrer rotation speed. For the step wise stirring methods, it is stated that the number of incremental steps needs to be sufficiently large such that each progressive step is correlated with the previous in order to satisfy the Nyquist sampling theorem – something which is counterintuitive to the normal nature of the RCs' operation. For the continuous stirring methods, the stirrers can simply be rotated at a given (constant) speed and the measurement samples obtained.

Once the measurement samples are obtained, the Doppler spectrum $D(f,\rho)$ can be deduced from Equation 2.63 [31].

$$D(f,\rho) = H(f,\rho)H^*(f,\rho) = |H(f,\rho)|^2 \qquad (2.63)$$

where $H(f,\rho)$ is the Fourier transform of the channel transfer function $H(f,t)$ with respect to time t and $\rho =$ Doppler frequency. It was shown in Ref. [31] that the step wise channel transfer function $H(f,t) = S_{21}(f,t)$ from time versus stirrer speed relations.

Instead of evaluating the Doppler spread at different thresholds, [31] introduced an RMS quantity which avoids the ambiguity that the different thresholds can pose. The RMS Doppler bandwidth (ρ_{RMS}) at a certain frequency f_o can be deduced according to Equation 2.64 [31].

$$\rho_{RMS} = \left[\frac{\int \rho^2 D(f_o,\rho)d\rho}{\int D(f_o,\rho)d\rho} \right]^{1/2} \qquad (2.64)$$

Figure 2.37 shows the Doppler frequencies measured in the University of Liverpool RC using continuous stirring methods. Three results are presented that show the Doppler shifts evident from three different stirrer speeds. Presented alongside to compare is a theoretical deduction of the Doppler frequency f_D calculated from Equation 2.65.

$$f_D = \frac{v}{\lambda} \qquad (2.65)$$

where $v = (2\pi r/T)$, $r =$ radius of the paddles $= 0.9\,m$ and T is the time taken for one complete rotation. It is stated in Ref. [31] that the right hand side of (2.65) must be multiplied by a factor of 2 when using this theory to predict the Doppler shifts from

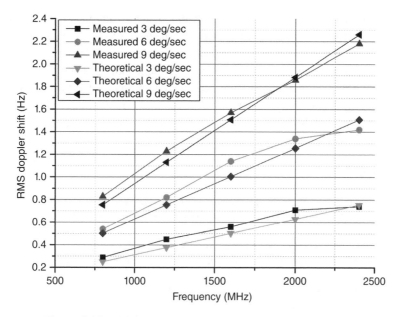

Figure 2.37   RMS Doppler shifts (Hz) from 800 to 2400 MHz.

Table 2.5   Doppler stirrer rotation times.

Indicated speed (deg/second)	Corresponding time (Second)
3	120
6	60
9	40

mechanical stirring paddles, since the paddles will reflect waves like a radar meaning that the Doppler shifts will be twice than observed from simply moving a transmitter or receiver in the same way.

For clarity, the timescale for one complete rotation of the mechanical stirring paddles in Figure 2.37 are stated in Table 2.5.

2.9.4   Summary

In this chapter, we have discussed some of the important unifying theories that uphold and support the RC as a measurement facility. Careful examinations on the nature of the electromagnetic fields inside the enclosure, stirring methods to 'stir' the electro-magnetic fields, plane wave AoAs, average mode bandwidths, Quality factors and the propagation channels statistical forms have been undertaken, and the focus has been on how each of these quantities can influence any practical measurement that is undertaken. In addition, channel sounding parameters have been introduced and discussed with some results also presented. The collation of the detailed theory presented in this chapter will enable the reader to acquire a detailed understanding of the operation of the RC, such that any practical undertaking can now be completed with underpinning and detailed theoretical knowledge to support. This will ensure confidence in any selected measurement parameters and aid in the accuracy of any measured results.

References

[1] R. F. Harrington, *Time-Harmonic Electromagnetic Fields*, New York: McGraw-Hill, 1961.

[2] C. A. Balanis, *Advanced Engineering Electromagnetics*, New York: John Wiley & Sons, Inc., 1989.

[3] J. D. Jackson, *Classical Electrodynamics*, 3rd ed.: New York: John Wiley & Sons, Inc., 1999.

[4] J. D. Kraus, *Electromagnetics*, 4th ed.: New York: McGraw Hill, 1991.

[5] D. A. Hill, *Electromagnetic Fields in Cavities: Deterministic and Statistical Theories*, New York: John Wiley & Sons, Inc., 2009.

[6] Y. Huang, 'The Investigation of Chambers for Electromagnetic Systems', D Phil Thesis, Department of Engineering Science, University of Oxford, 1993.

[7] C. L. Holloway, H. A. Shah, R. J., Pirkl, W. F. Young, D. A. Hill and J. Ladbury, 'Reverberation chamber techniques for determining the radiation and total radiation efficiency of antennas', *IEEE Transactions on Antennas and Propagation*, vol. 60, pp. 1758–1770, 2012.

[8] D. A. Hill, 'Boundary fields in reverberation chambers', *IEEE Transactions on Electromagnetic Compatibility*, vol. 47, pp. 281–290, 2005.

[9] P. S. Kildal and C. Carlsson, 'Detection of a polarization imbalance in reverberation chambers and how to remove it by polarization stirring when measuring antenna efficiencies', *Microwave & Optical Technology Letters*, vol. 34, pp. 145–149, 2002.

[10] P. S. Kildal, X. Chen, C. Orlenius, M. Franzen and C. L. Patane, 'Characterisation of reverberation chambers for OTA measurements of wireless devices: Physical formulation of channel matrix and new uncertainty formula', *IEEE Transactions on Antennas and Propagation*, vol. 60, pp. 3875–3891, 2012.

[11] K. Rosengren, P. S. Kildal, C. Carlsson and J. Carlsson, 'Characterisation of antennas for mobile and wireless terminals in reverberation chambers: Improved accuracy by platform stirring', *Microwave & Optical Technology Letters*, vol. 30, pp. 391–397, 2001.

[12] K. Rosengren and P. S. Kildal, 'Study of distributions of modes and plane waves in reverberation chambers for the characterisation of antennas in a multipath environment', *Microwave & Optical Technology Letters*, vol. 30, pp. 386–391, 2001.

[13] P. S. Kildal, *Foundations of Antennas: A Unified Approach*, Sweden: Studentlitteratur, 2000.

[14] S. J. Boyes, Y. Huang and N. Khiabani, 'Assessment of UWB antenna efficiency repeatability using reverberation chambers', *2010 IEEE International Conference on Ultra-Wideband (ICUWB)*, vol. 1, IEEE, 20–23 September 2010, Nanjing, pp. 1–4.

[15] BS EN 61000-4-21:2011, 'Electromagnetic Compatibility (EMC) testing and measurement techniques: Reverberation chamber test methods', 2011.

[16] X. Chen and P. S. Kildal, 'Theoretical derivation and measurement of the relationship between coherence bandwidth and RMS delay spread in reverberation chambers', *3rd European Conference on Antennas and Propagation, 2009.* EuCAP 2009, IEEE, 23–27 March 2009, Berlin, pp. 2687–2690.

[17] X. Chen, P. S. Kildal and L. Sz-Hau, 'Estimation of average Rician K factor and average mode bandwidth in loaded reverberation chamber', *IEEE Antennas and Wireless Propagation Letters*, vol. 10, pp. 1437–1440, 2011.

[18] P. Corona, G. Latmiral and E. Paolini, 'Performance and analysis of a reverberating enclosure with variable geometry', *IEEE Transactions on Electromagnetic Compatibility*, vol. 22, pp. 2–5, 1980.

[19] D. A. Hill, M. T. Ma, A. R. Ondrejka, B. F. Riddle, M. L. Crawford and R. T. Johnk, 'Aperture excitation of electrically large, lossy cavities', *IEEE Transactions on Electromagnetic Compatibility*, vol. 36, pp. 169–178, 1994.

[20] U. Carlberg, P. S. Kildal, A. Wolfgang, O. Sotoudeh and C. Orlenius, 'Calculated and measured absorption cross sections of lossy objects in reverberation chamber', *IEEE transactions on Electromagnetic Compatibility*, vol. 46, pp. 146–154, 2004.

[21] P. S. Kildal, C. Orlenius, J. Carlsson, U. Carlberg, K. Karlsson and M. Franzen, 'Designing reverberation chambers for measurements of small antennas and wireless terminals: Accuracy, frequency resolution, lowest frequency of operation, loading and shielding of chamber', *First European Conference on Antennas and Propagation, 2006*. EuCAP 2006, IEEE, 6–10 November 2006, Nice, pp. 1–6.

[22] W. C. Jakes, *Microwave Mobile Communications*, New York: John Wiley & Sons, Inc., 1974.

[23] J. G. Kostas and B. Boverie, 'Statistical model for a mode-stirred chamber', *IEEE Transactions on Electromagnetic Compatibility*, vol. 33, pp. 366–370, 1991.

[24] L. C. Andrews and R. L. Phillips, *Mathematical Techniques for Engineers and Scientists*, Bellingham: SPIE Press, 2003.

[25] F. J. Massey, 'The Kolmogorov-Smirnov test for goodness of fit', *Journal of the American Statistical Association*, vol. 46, pp. 68–78, 1951.

[26] H. W. Lilliefors, 'On the Kolmogorov-Smirnov test for normality with mean and variance unknown', *Journal of the American Statistical Association*, vol. 62, pp. 399–402, 1967.

[27] H. W. Lilliefors, 'On the Kolmogorov-Smirnov test for the exponential distribution with mean unknown', *Journal of the American Statistical Association*, vol. 64, pp. 387–389, 1969.

[28] C. L. Holloway, D. A. Hill, J. M. Ladbury, P. F. Wilson, G. Koepke and J. Coder, 'On the use of reverberation chambers to simulate a Rician radio environment for the testing of wireless devices', *IEEE Transactions on Antennas and Propagation*, vol. 54, pp. 3167–3177, 2006.

[29] P. S. Kildal, S. H. Lai and X. Chen, 'Direct coupling as a residual error contribution during OTA measurements of wireless devices in reverberation chamber', *2009 IEEE Antennas and Propagation Society International Symposium and Usnc/Ursi National Radio Science Meeting*, vols 1–6, IEEE, June 2009, pp. 1428–1431.

[30] X. Chen, P. S. Kildal, C. Orlenius and J. Carlsson, 'Channel sounding of loaded reverberation chamber for over the air testing of wireless devices – Coherence bandwidth vs average mode bandwidth and delay spread', *IEEE Antennas and Wireless Propagation Letters*, vol. 8, pp. 678–681, 2009.

[31] K. Karlsson, X. Chen, P. S. Kildal and J. Carlsson, 'Doppler spread in reverberation chambers predicted from measurements during step wise stationery stirring', *IEEE Antennas and Wireless Propagation Letters*, vol. 9, pp. 497–500, 2010.

3

Mechanical Stirrer Designs and Chamber Performance Evaluation

In this chapter we will discuss a technique and go through the process of designing mechanical stirring paddles for use in Reverberation Chambers (RCs). The overall aim of this chapter is to detail a method of how the paddle design process can be accomplished. The paddle design in this case looks to improve chamber performance down towards lower modal numbers, where the performance of all chambers will begin to degrade.

In order to assess the performance of any stirring process, the RC must be practically assessed and validated. In this chapter, we will also present detailed procedures, guidance and mathematical equations to explain and show how the practical performance of an RC is validated in a manner that is both robust and practically useful. This will aid any practitioner who uses RCs to undertake the validation process with confidence, and also to yield useful, accurate information about how their chamber performs.

3.1 Introduction

As was established in Chapter 2, the field distribution in an 'un-empty' cavity, outside of the source area, is the superposition of all Transverse Electric (TE) and Transverse Magnetic (TM) cavity modes. Theoretically it has been shown that fields in the chamber are formed as a result of standing waves that have a sine and cosine dependence. Furthermore, we have also established that mechanical stirring is one of the key tools

Reverberation Chambers: Theory and Applications to EMC and Antenna Measurements, First Edition.
Stephen J. Boyes and Yi Huang.
© 2016 John Wiley & Sons, Ltd. Published 2016 by John Wiley & Sons, Ltd.

that is used in RCs to render this field distribution statistically homogeneous and isotropic on average, and thus make the environment suitable for Electromagnetic Compatibility (EMC) and antenna measurements.

In any given RC it is widely acknowledged that the chambers' performance will improve with increasing frequency. The reasons for this are:

1. For an increase in frequency there is an increase in the number of modes governed approximately by Weyl's law, giving the amount of modes available proportional to the third power of frequency and the chamber volume.
2. Different modes need to be excited in order to promote a change in the field distribution.

In addition to points (1) and (2), it can also be argued that the performance in the chamber is also a result of the mechanical stirring paddles in the chamber being electrically large enough to be able to adequately stir the modes (i.e. the paddles can change the boundary conditions sufficiently such that it causes the resonant frequencies of the modes to shift).

Conversely, as previously shown, towards lower frequencies in a given chamber fewer modes will exist – in terms of the frequency domain spectrum, the individual mode resonant frequencies are spaced further apart making it more difficult to excite and stir different modes. It is apparent that at larger wavelengths, the electrical size of the paddles becomes smaller, thereby diminishing their ability to adequately interact with and stir the modes.

In terms of the frequency domain spectrum, Figure 3.1 illustrates the modal deficiencies at lower frequencies in the University of Liverpool RC. Each vertical line that is shown represents a mode.

As Figure 3.1 shows, towards lower frequencies it will become more difficult to excite different modes and promote a sufficient change in the field distribution. The bell-shaped curve from Figure 3.1 serves to illustrate the 'average mode bandwidth' which is defined as: *The bandwidth over which the excited power in a particular cavity mode with resonance frequency f_0 is larger than half the excited power at f_0.* Despite the narrow width it potentially proves that many modes can still be excited at high frequencies.

Upon accepting the prior rationale, the question therefore is: What can one do about this at lower frequency?

In terms of the modal structure, one could be tempted to severely load the chamber to broaden the average mode bandwidth. However, having too large an average mode bandwidth will mean that the mechanical stirring process will struggle to sufficiently change the modal structure enough to excite different modes about the large modal bandwidth. The consequence of this is that the field distributions would not sufficiently change and a large standard deviation in any measurements would ensue.

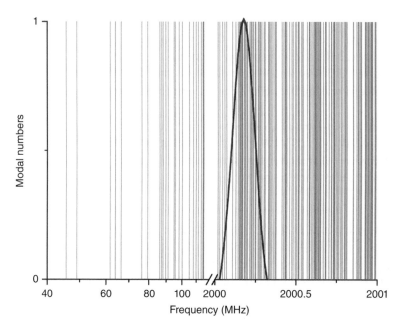

Figure 3.1 Modal density in low and high frequency domain.

In addition, by loading the chamber with large amounts of anechoic absorber (for example), this would dampen resonant modes and the subsequent Angle of Arrival (AoA) of plane waves from given directions, resulting in an increase in the proportion of direct power in the chamber (increased Rician K factor). Neither of these consequences is necessarily desirable.

One option could be to just simply increase the size of the chamber such that at a desired frequency of operation there are plenty of available modes. This is true, but the chamber would always face the same problem as the frequency decreased.

In terms of the mechanical stirring paddles themselves, different researchers have looked into the issue of design. At the time of writing, there does not appear to be any standardised format as to how the paddles should be designed or shaped; however, a few 'rules of thumb' do exist. Generally, the stirrer should be electrically large at the operating frequency [1]. Furthermore, since very large stirrer designs are liable to occupy precious space inside the chamber, two or more stirrers are often used in practice to provide equivalent or better performance [2].

Practically, the effect of the stirrer angle, height and width (aspect ratio) was studied in Ref. [3] as a function of chamber Q factor, field uniformity, number of excited modes and stirrer efficiency (number of independent samples). It was concluded that multiple stirrers performed best, and a 90° plate angle coupled with a minimum 80 cm height in conjunction with the use of 1λ reflectors were offered as optimised dimensions in that case.

Wellander, Lunden and Backstrom undertook an experimental and mathematical modelling investigation in Ref. [4]. It was concluded here that the efficiency in which the stirrer operates can be greatly improved by increasing its diameter. However, increasing a stirrer's physical diameter will result in that stirrer occupying more area inside the chamber, and consequently the working volume in the chamber will be decreased. This is not desired, and hence as an alternative, it was stated that the stirrers could be re-designed by varying the aspect ratios (height/width) to occupy less volume [4].

Arnaut presented a mathematical formulation in Ref. [5] based on the effect of size, orientation and eccentricity of mode stirrers. It was found that for a sufficiently large radius of rotation, the eccentricity of a stirrers' design had a greater effect on the stirrers' performance than the effect of the aspect ratio or relative orientation. This conclusion agreed with Clegg's earlier research in Ref. [6] who proposed a randomised plate stirrer based on the work of a genetic algorithm in conjunction with a Transmission Line Matrix (TLM) software; the TLM approach is also being proposed in Ref. [7].

Reviewing the aforementioned published works, the one common aspect that they all have in common (excluding [1] and [2]) is that the stirrer designs have all been founded upon the use of solid metallic plates. In this chapter we are going to have a look into the physical paddle design to see if any practical improvements can be realised. The simple motivation with this design process is to see if any improvements in performance can be realised down towards lower modal densities in a broadband manner, to improve upon degraded chamber performance.

This chapter will first discuss the design methodology and explain the theoretical principles about the paddle design that will be proposed. From here we will detail the numerical analysis that can be used to aid in the paddle design process. A practical validation phase will also be presented where we will show in detail how the performance of an RC can be robustly assessed.

3.2 Paddle Design Methodology

With regards to the paddle structure, the design methodology we will use proposes the creation of various cuts across the paddle structure as opposed to leaving the paddles as solid metallic sheets. Figure 3.2a and b illustrates this concept.

The theory supporting such a modification is based on the concept of meander line theory. As mentioned in the introduction (Section 3.1), it is accepted that increasing the diameter of a stirrer will improve the stirrers' performance at lower frequencies. From the meander line theory, it is well known that this technique can enable the lowering of the resonant capability of an antenna structure without increasing its overall physical size [8].

(a)   (b)

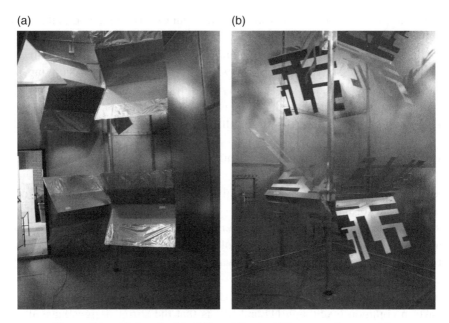

Figure 3.2   (a) Standard (solid) paddles and (b) cuts made on paddle.

Therefore, in this case, the paddle structure is to be treated in a similar manner as an antenna. The meander line principle has been adopted as a means of increasing the electrical size of the structure while leaving the physical dimensions of the paddle the same. Hence, the cuts on the paddle are designed to increase the current path length on the plates, with the current being induced whenever a plane wave comes into contact.

The induced current on a paddle is stated to be at maximum when the stirrer is resonant [7]; this is why the resonant capability of the paddle is aimed to be driven lower. With a longer current path length, the stirrer can be resonant at lower frequencies. If the stirrer has lower resonant capabilities, it has the potential to interact with and 'stir' modes at lower frequencies where standard paddles will not perform that well.

Another motivation with regards to the design concerns the statement of eccentricity in Ref. [5]. The eccentricity can be formed in this case as a result of all the varying cuts on the paddle having different sizes. Hence as this structure rotates, the various cuts and shapes will provide for an extreme eccentric volume of rotation, which is aimed at improving performance.

Some questions are probably prevalent now:

1. How has such a design come about?
2. What evidence exists to suggest this approach works and is better performing?

These two points will be conclusively addressed in Section 3.3. Beforehand, a few comments on the numerical method of analysis are warranted.

The design methodology has been performed using an Eigenanalysis in conjunction with a given physical stirrer design. Since it is well known (and has been proved in Chapter 2) that the field distribution in a given cavity can be written in terms of cavity modes [9], and a shift in the resonant frequency of a mode implies a change in the field distribution [7], then the following can be applied.

The proposed Eigenanalysis technique in this chapter is based on the use of a spherical cavity. It is accepted that spherical cavities do not necessarily make good RCs owing to the presence of multiple degenerate modes, but the spherical cavity approach does have advantages in this case:

- It can save time and computational resources as the stirrer paddles do not need to be rotated as the cavity is entirely symmetrical.
- Rectangular cavities would require the paddles to be rotated and any results obtained would be a function of the number of stirrer increments.
- All designs presented here are assessed under the same conditions so the relative performance can be compared.

Therefore, by use of this approach one should:

1. Obtain the eigenfrequencies in an empty cavity (λf_{empty})
2. Introduce each subsequent design into the cavity and calculate how the modes are permuted (λf_{loaded}).
3. Calculate the eigenfrequency shifts ($\Delta \lambda f$) according to (3.1).

$$\Delta \lambda f = \left| \lambda f_{empty} - \lambda f_{loaded} \right| \qquad (3.1)$$

Simply, the design that permutes the eigenfrequencies by the greatest amount is the optimum choice. This is consistent with Wu and Chang [7] who stated that the key mechanism behind an effective stirrer lies in its ability to shift the eigenfrequencies.

3.3 Numerical Analysis

The numerical analysis has been performed using a well-known commercial software, CST Microwave Studio [10]. Figure 3.3 serves to illustrate the set-up used in this investigation.

From Figure 3.3, the internal volume of this cavity is chosen as $2.144 \, m^3$ (radius = 0.8 m). This selection was made as a trade-off between the total number of modes available in the cavity and the overall computational time and resources that are required in order to yield a solution.

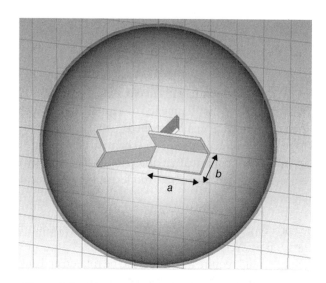

Figure 3.3   Numerical set-up showing standard paddles.

Three plates are seen to be co-joined about the centre of the spherical cavity – the co-joined aspect has been configured to be similar to the real physical designs in the existing chamber (see set-up in Figure 3.2). It is entirely conceivable however that the plate construction can easily be designed around a pre-existing layout in any given chamber if that was to be retained.

The dimensions of the numerical plates in this investigation have been configured as width $(a) = 0.4$ m and (b) height $= 0.35$ m. The reason for this selection was because the aspect ratio $(b/a) = 0.875$ is equal to the aspect ratio of one of the paddle sets in the real physical chamber. Hence, the numerical set-up is again based as much as possible on physical reality, which can be easily configured for different plate dimensions and layouts.

The simulated cavity is constructed from aluminium, 1-cm thick, and the material filling the cavity is air (free space). The boundary conditions selected for the numerical investigation was $E = 0$; that is, the field converges to zero at the cavity walls. The reason for this selection is that this is the typical boundary condition that is implied in the real physical chamber.

In terms of the total lengths of the meander line cuts, it was decided to make this length proportional to the wavelength of where the first three overlapping modes occur. In the simulation, this corresponds to 0.986 m $= 304$ MHz; when scaled up to the real physical design, this translates to 2.6 m $= 115$ MHz – that is, significantly larger than the original paddle dimensions (width $= 1$ m and heights $= 0.58$–0.9 m) in order to lower the resonant capability.

The reasons for this selection can be given as follows:

1. The wavelength selected for the real physical design corresponds to a frequency where at least one mode per megahertz (mode density) exists. A lower frequency was not chosen because the modal structure would simply be too sparse.
2. The length of the cuts at this range provided a degree of flexibility with regards to how the cuts could be structured. Longer length cuts would inhibit this flexibility somewhat.
3. A higher frequency wasn't selected as the bound on the resonant capability of the paddle structure was envisaged to be pushed to the maximum that could possibly be achieved.

There is conceivable scope within this approach to be useful to any paddle type designs, chambers and different frequencies if so desired.

3.3.1   Effect of the Number of Cuts

As stated previously, we are looking here at establishing what type of design can perform better at lower modal numbers. As a first step, the total number of cuts on the paddle is numerically investigated. Table 3.1 details the parameters used in this investigation.

Figure 3.4a and b provides a visual representation of the numerical models in this investigation and Figure 3.5 details the numerical Eigenshifts result.

From Figure 3.5 it can be seen that in this case, a lower number of cuts performs better than a larger number. Computationally speaking, this result is also advantageous as it takes fewer resources to yield a solution for a lower number of cuts. Although there are areas where the eigenfrequencies of the modes are significantly shifted from their original (empty cavity) resonant frequencies, there are also areas where little or no eigenfrequency shifts are present. In practice, a broadband response would be advantageous instead of just isolated frequencies.

Table 3.1   Parameters for investigation on number of cuts.

Parameter	Description
Simulation set-up	Rectangular cuts
	90° plate angle
	Cuts all periodic in nature
	Paddles all the same dimension

(a)   (b)

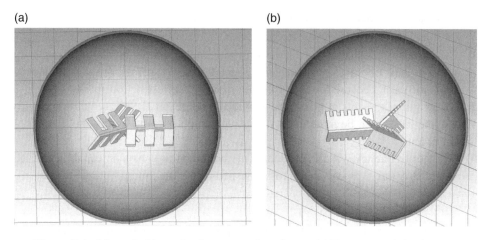

Figure 3.4   Numerical investigation on number of cuts. (a) Two cuts and (b) six cuts.

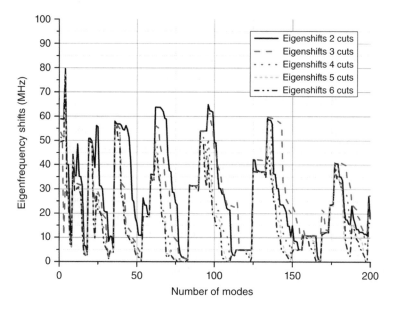

Figure 3.5   Eigenfrequency shifts (MHz) vs number of cuts.

3.3.2   *Effect of the Periodicity of the Cuts*

As a second step, the effect of periodicity of the cuts is investigated; that is, should the cuts be periodic or non-periodic in nature? The length, width and non-symmetrical designs are looked at here. What this investigation is trying to establish is the effect of the frequency response with respect to the nature of the designs. Each specific

Table 3.2 Parameters for periodicity investigation.

Parameter	Description
Simulation set-up	Rectangular cuts
	90° plate angle
	Varied length of cuts only
	$3 \times (\lambda/2)$ cuts chosen:
	1. 0.37 m
	2. 0.24 m
	3. 0.177 m
	Paddles all the same dimension
	Mirror image design

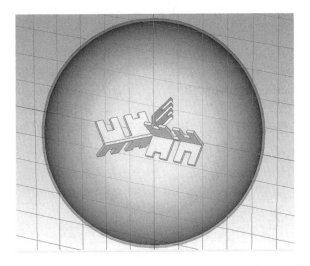

Figure 3.6   Non-periodic investigation 1, varying just length of cuts.

dimension of cut is considered to be a half wavelength multiple of the frequency (see Table 3.2) in which to enhance. Table 3.2 details the investigation parameters.

Figure 3.6 details the numerical set-up for this investigation and Figure 3.7 depicts the eigenfrequency shift results.

From Figure 3.7 it can be seen that specific enhancements are provided with the varying lengths of cuts as opposed to having all the cuts made periodically. However, the design is still not sufficient as it is not a desired broadband response.

For the next step, both the length and width of the cut was varied to assess the overall response. Table 3.2 details the main parameters, although the dimensional size of the cuts was varied slightly to reflect the changing circumstances. $3 \times (\lambda/2)$ cuts chosen: (1) 0.39 m, (2) 0.26 m and (3) 0.177 m.

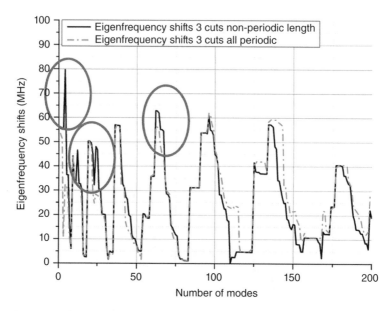

Figure 3.7   Eigenfrequency shifts (MHz) varying just length of cuts vs number of modes.

Figure 3.8   Non-periodic investigation 2, varying both length and width of cuts.

Figure 3.8 portrays the numerical set-up, while the eigenfrequency shift results can be seen in Figure 3.9.

From Figure 3.9, the response is seen to be receptive to the changing lengths and widths of the cuts to a certain extent; however, the design is still not broadband, and performance here is not satisfactory.

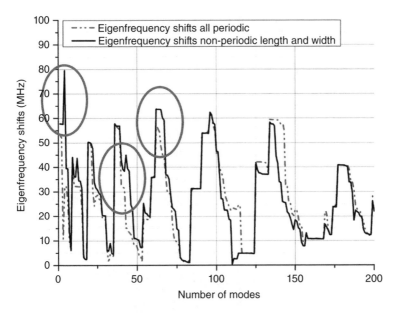

Figure 3.9 Eigenfrequency shifts (MHz) varying both length and width of cuts vs number of modes.

As a fourth step, the design was also varied such that the top and bottom of the plate sections were not a mirror image. The same parameters as in Figure 3.8 were applied. The results showed a marginal difference to those depicted in Figure 3.9, so for brevity they will be omitted here. In practice however, it is desired not to have rotational symmetry [5].

Therefore, retaining the varying length, width and non-mirrored design strategy, the shape of the cuts was investigated next.

3.3.3   Effect of the Shape of the Cuts

Three different meander line shapes are compared in this section; the rectangular shape which has already been disclosed, a triangular meander line and a normal mode helical shape. This investigation begins by considering a triangular meander line structure. Table 3.3 details the numerical parameters.

Figure 3.10 details the numerical model, while the eigenfrequency shifts can be viewed in Figure 3.11.

From Figure 3.11 it can be seen that the rectangular meander line exhibits a lower resonant capability and a better frequency response. The effect of varying the triangular cut angles was also investigated and was found to offer no significant improvement to the results seen in Figure 3.11.

Table 3.3   Parameters for effect of cut shape investigation 1.

Parameter	Description
Simulation set-up	Triangular cuts
	90° plate angle
	Varied length of cuts
	Cut angles periodic (40°)
	$3 \times (\lambda/2)$ cuts chosen:
	1. 0.3236 m
	2. 0.206 m
	3. 0.08 m
	Paddles all the same dimension
	Design not mirror image

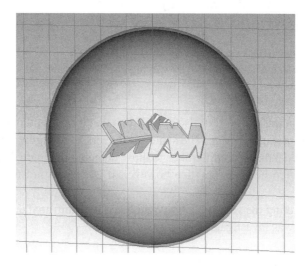

Figure 3.10   Effect of cut shape investigation 1: Triangular meander line.

So far, the investigation has determined that the rectangular meander line has a lower resonant capability than the triangular meander line – a fact also validated in Ref. [8] for wire monopole antennas, and that the overall frequency response can be improved for varying lengths and widths of cuts. The results seen so far however are still problematic in the sense that the improvements are only narrowband in nature; a broadband response is desired.

Furthermore, from a manufacturing perspective, the rectangular (straight edge) cuts would possibly be an easier proposition to create than varying angles and lengths of triangular cuts. Therefore, the decision was taken not to pursue the triangular

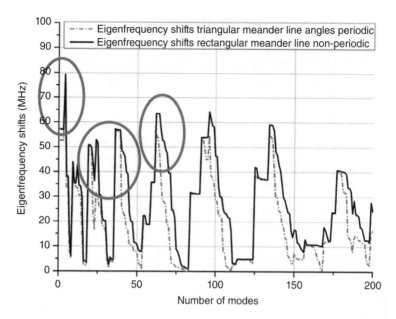

Figure 3.11 Eigenfrequency shifts (MHz) for rectangular and triangular meander line shapes vs number of modes.

Table 3.4 Parameters for effect of cut shape investigation 2.

Parameter	Description
Simulation set-up	Normal mode helix cut
	90° plate angle
	Same length of cuts: Period = 0.2465 m
	Paddles all the same dimension
	Design not mirror image

meander line theme further. The next investigation looked at a normal mode helical design to assess the performance of such a meander line shape. Table 3.4 details the numerical parameters.

Figure 3.12 details the numerical set-up, while the eigenfrequency shift results can be viewed in Figure 3.13.

From Figure 3.13 it is shown that the helix-type structure has greatly enhanced the resonant capabilities at lower modal numbers – in this range it is by far superior to the rectangular meander line-shaped cuts. However, the design is not broadband as the periodicity of the cuts is the same. To make the structure broadband, it is envisaged having many different $\lambda/2$-sized individual cuts on the paddle structure, such that in a given range, there is a cut comparable or nearly comparable in size in order

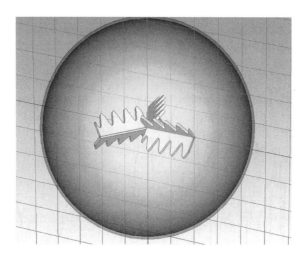

Figure 3.12   Effect of cut shape investigation 2: Helical meander line.

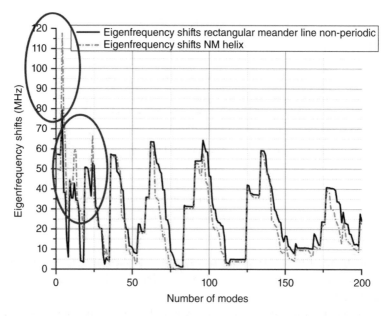

Figure 3.13   Eigenfrequency shifts (MHz) for rectangular and helical meander line shapes vs number of modes.

for the paddle to interact with a plane wave efficiently. Please see Appendix C for evidence to support this statement.

With the helix-shaped cuts, it was envisaged that difficulties would result in trying to fit many different-sized cuts onto one paddle structure. Although the helix-shape structure shows a degree of promise, the decision in this investigation was taken not to pursue the helix meander line for the aforementioned reasons.

Koch fractal-type shapes were also disregarded due to their complex nature of manufacture, although these types of cuts could have potentially improved performance.

3.3.4   Complex Nature of the Cuts

Since the design strategy looks to having many different $\lambda/2$-sized cuts on the paddle to offer a broadband characteristic, the application of having straight edge cuts would offer a degree of flexibility in structuring all these different cuts onto one paddle structure.

This investigation therefore looked into the complex nature of rectangular meander line cuts on the paddle structure; specifically, should the cuts be made in both a vertical and horizontal plane? The guiding theory supporting this move is as follows:

1. As previously stated, the more cuts available on the paddle is envisaged to offer a broadband characteristic – the vertical plane would be needed to fit all these cuts on the paddle.

Although the existing chamber has two principle sets of mechanical stirring paddles for vertical and horizontally polarised waves, when the chamber is employed for measurements, the transmitting antenna is typically directed straight at the vertically mounted stirrer paddles as seen in Figure 3.14.

Figure 3.14   Transmitting antenna location.

Table 3.5 Parameters for the complex nature of cuts.

Parameter	Description
Simulation set-up	Rectangular meander line design
	90° plate angle
	$10 \times \lambda/2$ cuts all different dimensions
	Paddles all the same dimension
	Design not mirror image
	Completely non-periodic

As discussed and detailed previously, polarisation stirring is employed. This is accomplished in our case by directing the transmitting antenna in two orthogonal orientations directly at the paddles. Therefore, cuts in both a vertical and horizontal plane on the paddle structure could theoretically serve both of these transmitting polarisations.

Other chamber designs exist, which employ multiple transmitting antennas inside that are orientated to excite the chamber with a combination of vertical and horizontally polarised waves, instead of relying on a single antenna as shown here. However, the process of creating cuts on the paddle should still serve a purpose as a generic way of increasing the electrical size of the paddles and hence improving performance at lower modal densities.

With respect to the prior theoretical rationale, Table 3.5 details the numerical parameters for this investigation.

Figures 3.15 and 3.16 depict the complex numerical model.

From Figure 3.16, the selections in the dimensions of cuts have been specifically made to provide a resonant capability in the rectangular designs where before there wasn't any. The selections are made simply to span a varying range of wavelengths, and the nature of how the cuts are orientated was made simply to be able to fit all cuts onto the metallic plate; this explains the complex nature of the design. Figure 3.17 portrays the numerical results for the complex design as compared with the (original) non-periodic rectangular meander line with only three varying cuts.

Analysing Figure 3.17, it can be seen that the complex nature of the design coupled with the use of varying specific length cuts grossly outperforms the simple-natured (original) design. The complex design here has a broadband frequency response that is desired. The numerical results shown in Figure 3.17 provides confidence to re-enforce the theory of specific $\lambda/2$ cuts, and provides evidence to assert that highly irregular-shaped mechanical stirring paddles can perform better in terms of their mode stirring performance.

Comparing the complex design to standard paddles with no cuts at all, as shown in Figure 3.18, we see the enhanced performance of the complex design strategy.

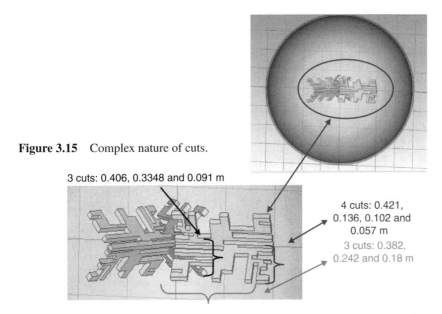

Figure 3.15   Complex nature of cuts.

3 cuts: 0.406, 0.3348 and 0.091 m

4 cuts: 0.421, 0.136, 0.102 and 0.057 m

3 cuts: 0.382, 0.242 and 0.18 m

Figure 3.16   Close-up of complex nature of cuts.

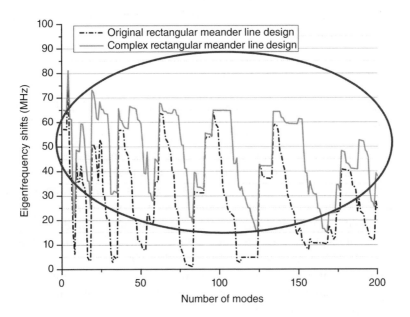

Figure 3.17   Eigenfrequency shifts (MHz) for complex and original rectangular meander line.

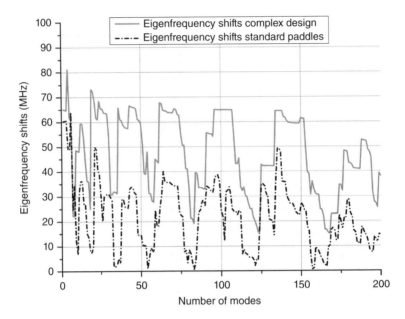

Figure 3.18 Eigenfrequency shifts (MHz) for complex and standard (no cuts) paddles.

With confidence established in the theoretical idea, the next step in the design process would be to attempt to improve the performance even further by looking at the lengths of each paddle structure.

3.3.5 Variation in Paddle Dimensions

Briefly referring back to the earlier statement concerning eccentricity and to the rationale presented in Ref. [5], and also to the effect of having rotational symmetry in Ref. [7], it has been known for a long time that rotational symmetry within an electrically large design and a concentric mode of operation would serve to inhibit randomness. Hence, in an RC, when rotational patterns are repeated, no additional stirring is gained [7].

To be theoretically precise, the effect of a stirrers scattering/diffusion performance should not be taken in isolation from the details and vicinity of its cavity boundary [5]. Hence, it has been stated that a small and simple stirrer inside a highly complex corrugated cavity might produce a statistically more uniform field distribution than a large complex stirrer inside a cavity with simple smooth walls [5].

In this case, since the walls of the chamber, and indeed the vast majority of chambers, are rectangular and smooth/metallic in nature, the statement concerning the rotational symmetry and eccentricity is applied.

In this investigation the dimensions of each of the three numerical paddle sets are varied in an attempt to assess the effect of employing no rotational symmetry and an extreme eccentric design.

Two separate investigations are presented:

1. One of the paddle dimensions retains its original dimensions; the other two paddle sets are scaled 1.25 and 1.5 times *in height only*.
2. One of the paddle dimensions retains its original dimensions; the other two paddle sets are scaled 1.5 and 1.75 times *in height only*.

Table 3.6 details the selections made in this regard, while Figure 3.19 depicts the typical numerical model.

Figure 3.20 displays the numerical results comparing the effect of changing the paddle dimensions for the complex design.

Table 3.6   Parameters for the variation in paddle dimensions.

Parameter	Description
Simulation set-up	Rectangular meander line design
	90° plate angle
	$10 \times \lambda/2$ cuts all different dimensions
	Design not mirror image
	Completely non-periodic
	$1 \times$ paddle original size
	$1 \times$ paddle scaled 1.25 and 1.5 times in height
	$1 \times$ paddle scaled 1.5 times and 1.75 times in height

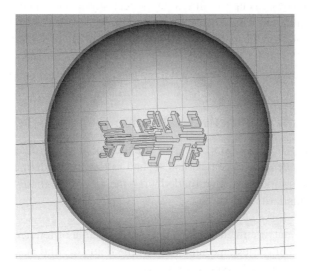

Figure 3.19   Complex cuts with differing paddle heights.

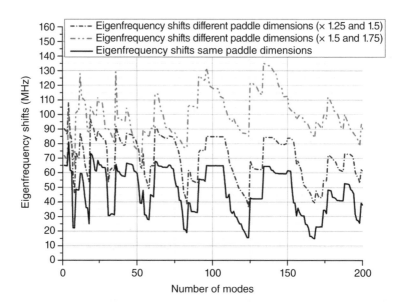

Figure 3.20 Eigenfrequency shifts (MHz) for complex paddles with different dimensions.

From Figure 3.20, we see that as the electrical size of the paddles increase, it better serves to permute the modes in the cavity. In addition, because each of the paddle sets is different in dimension, the cuts made on those paddles are also scaled accordingly – this means that a wider broadband nature can be provided for as there are more different-sized cuts on the entire paddle structure.

In practice, there are dimensional constraints that in this case had to be adhered to. When scaling the designs up to full size in the real physical chamber, the 1.5 and 1.75 times increase in height would be prohibitive in the University of Liverpool RC as the paddles would catch on the walls, given the proximity of where the paddles are situated. However, the 1.25 and 1.5 times increase in height do allow for adequate clearance, so the results here serve to illustrate what will be taken into practice validation section.

3.4 Comments on Practical Validation

Conventionally, when assessing the performance of a given chamber, one would typically refer to the relevant standards [11] and follow the guidelines issued to determine the chambers' performance by deducing the electromagnetic field uniformity. Typically, the standards advocate:

1. Measuring the maximum electric (E) field value at eight different chamber locations for three different Cartesian orientations of the field (x, y and z).
2. Use logarithmic spaced frequency points.

3. Number of measurement samples (stirrer steps) proposed = 12 (minimum).
4. Calculate the chamber performance by assessing the standard deviation in the maximum recorded electric field values – for each x, y and z orientation and when all the orientations are combined.

This procedure is typically employed for the EMC measurement uncertainty characterisation; but for antenna measurements, there are further chamber performance quantities that are desirable to know. For example, we have previously spoken about the requirement for minimising any differences in the received power as a function of the receive antennas polarisation through the use of polarisation stirring. Thus for antenna measurements, knowing what polarisation imbalance exists would be useful. Furthermore, knowing what proportion of power can flow direct to a receiver with minimal chamber/stirrer interaction is also useful to know to assess how the chamber performs and what impact this can have on any results obtained.

These differences may be more pronounced when conducting measurements with antennas as opposed to specific pieces of equipment undergoing EMC tests that do not have any performance merits linked to polarisation. Therefore, a robust method of how the uncertainty in a given chamber should be detailed that reflects its use potentially across both domains (for EMC and/or antenna measurements).

In this subsection, we will discuss and detail a more robust means of practical chamber characterisation that provides a more detailed account of how an RC performs – this is because the chamber is also used to conduct a wide series of antenna measurements in addition to EMC. If we assess current literature, it can be seen that other researchers [4, 12–15] are also choosing to widen the parameter scope when assessing their chamber measurement uncertainty.

In this work, the assessment of using a standard deviation from different measurement locations in the chamber will be retained, but crucially the standard deviation will be calculated as the deviation in the average power transfer function at different measurement locations in the chamber – using all measured samples, instead of just the (minimum of) 12 maximum electric field values [11], which is consistent with how antenna measurements are conducted in the RC.

Furthermore, looking back at the theory from Chapter 2, the following three parameters will also be assessed:

1. The non-line of sight (NLoS) number of independent samples as a function of frequency.
2. The measured polarisation imbalance that exists as a function of frequency.
3. The proportion of direct power (Rician K factor) as a function of frequency.

The standard deviation and the abovementioned points (1–3) form more of the components that can contribute to the uncertainty in antenna measurements. By this, these procedures should provide more of a complete indicator of the chambers' true performance.

3.5 Measurement Parameters for Validation

Table 3.7 details the parameters applied for this practical investigation. With respect to the parameters detailed, six receiver locations have been selected as a trade-off between the measurement accuracy and the overall measurement time needed to complete each measured sequence. Also, since each receiver location should be separated by at approximately $\lambda/2$ so that each location is independent (uncorrelated); at lower frequency in the University of Liverpool RC, this meant that any additional receiver locations could not be included as they would be too closely spaced. When performing such procedures, this is one aspect to check carefully prior to measurement.

The total number of measurement samples has been chosen partly through experience of using the particular chamber for measurements. This will become apparent in more practically orientated chapters to follow. It is conceivable however that this could vary depending on a users' own experience of using his/her chamber. This parameter could and possibly should also be varied in an individual chamber to chart the effect of the measurement uncertainty vs the overall time required to conduct each measurement sequence.

It is beyond the scope of this book to present generic measurement parameters for all chambers, the purpose here is to discuss and document the uncertainty procedures and relevant mathematical equations.

Table 3.7 Parameters for the practical investigation.

Parameter	Description
Frequency (MHz)	100–1000
Number of frequency data points	801
Number of independent receiver locations	6
Stirring sequences	Three degree mechanical stirring
	Polarisation stirring
	18 MHz frequency stirring
Receive antenna orientations	Three (x, y and z orientation)
Total number of measured samples per receiver location (for all orientations)	714
Miscellaneous	Source power $= -7\,$dB m
	RX antenna $=$ Log-Periodic (HL 223)
	TX antenna $=$ Homemade Vivaldi
	Chamber loading (when used) $=$ four pieces of anechoic absorber, two each in the back two corners of chamber
	Full 2 port calibration performed

3.6   Measurement Results

This section is subdivided into two separate sections to assess the unloaded and loaded chamber performance separately. In each section, a direct comparison will be made from standard stirrer paddles (no cuts at all) to the new stirrer designs to assess any improvement in practical performance.

The new stirrer designs from the numerical model have simply been scaled up in size to match the dimensions of the original stirrer paddles in the chamber – no changes whatsoever in the overall dimensions has been made between the respective stirrer sets so that the effect of the meander line cuts alone can be conclusively charted.

Please note that for the concept validation, only the vertically mounted stirrer paddles have been subject to modification. The paddles designed for horizontally polarised waves have remained as solid plates with no modifications whatsoever. For the vertically mounted paddles, six separate paddles exist each with different height dimensions. For completeness these are listed in Table 3.8.

We will now go through the practical process and the procedures of how the uncertainty in an RC can be deduced.

3.6.1   Standard vs New Designs: Unloaded Chamber Uncertainty

With respect to Table 3.7, the first measurement assessment concerns the standard deviations. As previously stated, this is the standard deviation in the average power transfer function from six different receiver locations in the chamber.

The specific receiver locations (constant throughout) are detailed as follows:

1. Pos 1 = 1.15 m from left wall, 1.1 m back from stirrers
2. Pos 2 = 1.15 m from right wall, 1.1 m back from stirrers
3. Pos 3 = 1.15 m from left wall, 1 m back from Pos 1
4. Pos 4 = 1.15 m from right wall, 1 m back from Pos 2
5. Pos 5 = 1.15 m from left wall, 1.1 m from back wall
6. Pos 6 = 1.15 m from right wall, 1.1 m from back wall

Table 3.8   Paddle dimensions.

Parameter	Description
Paddle set 1 (height × width in metres)	0.71 × 1
Paddle set 2 (height × width in metres)	0.85 × 1
Paddle set 3 (height × width in metres)	0.9 × 1
Paddle set 4 (height × width in metres)	0.67 × 1
Paddle set 5 (height × width in metres)	0.77 × 1
Paddle set 6 (height × width in metres)	0.62 × 1

Figure 3.21 Standard deviation measurement set-up.

Figure 3.21 shows the typical set-up in the measurements, depicting the new stirrer designs and the receive antenna in position 1, orientated in the x direction.

The average power transfer function (P) was deduced from (Eq. 3.2) for each receive antenna orientation (j) and at each specific location (i) using 3 degree mechanical stirring and polarisation stirring. Thus:

$$P_{i\_j} = \frac{\left\langle |S_{21}|^2 \right\rangle}{\left(1 - |S_{11}|^2\right)\left(1 - |S_{22}|^2\right)} \quad (3.2)$$

where $i = 1$, 2, 3, 4, 5 and 6 and $j = 1$, 2 and 3. Hence, at the end of the measurement sequences, 18 separate average power transfer function measurements should exist per frequency point. An average is then formed, comprising all $N = 18$ measurements:

$$P_{AV} = \frac{1}{N}\sum P_{i,j} \quad (3.3)$$

where this average should represent a good estimation of the true expected value [12].

The standard deviation (σ) is then calculated using (3.4).

$$\sigma = \sqrt{\frac{\sum_{i=1}^{6}\sum_{j=1}^{3}\left\{\left(P_{i,j}\right)-\left(P_{AV}\right)\right\}^2}{N}} \qquad (3.4)$$

In this case $N=18$ instead of $N-1$ has been applied because the entire dataset population (population standard deviation) has been used in the calculations [16] – that is, every single measurement value as opposed to just the selected maximums (sample standard deviation) in Ref. [10]. Finally, the standard deviation in dB scale is expressed relative to the mean and deduced using (3.5).

$$\sigma\left(dB\right) = 10\log_{10}\left\{\frac{\sigma+\left(P_{AV}\right)}{\left(P_{AV}\right)}\right\} \qquad (3.5)$$

Figure 3.22 details the measured standard deviation comparison between the standard paddles (no cuts) and the new modified paddles as a function of frequency in the unloaded chamber. It can be seen that up to approximately 380 MHz the new stirrer design outperforms the conventional (solid plate) paddle design. In the 100–200 MHz range, the improvement in performance is seen to be in the order of 0.5 dB, which beyond 200 MHz decreases to around 0.15 dB.

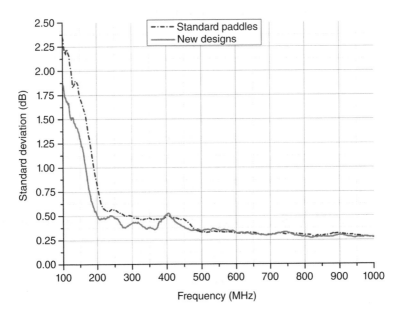

Figure 3.22 Standard deviation comparison in unloaded chamber.

Towards higher frequencies the performance of both paddle sets is seen to be comparable, proving that up to 1000 MHz, the cuts on the new paddle do not seem to have affected the high frequency performance of the chamber.

The results seen so far would appear to practically validate the theoretical concept; however three more performance indicators need to be compared in order to derive more confidence in the performance merits of the new designs.

The polarisation imbalance can be detailed as follows, after consultation with Ref. [11]. Considering just the x orientated measurements, a power transfer function can be established from the mechanical stirring and polarisation stirring schemes to yield (3.6).

$$P_{i,x} = \frac{\left\langle |S_{21,x}|^2 \right\rangle}{\left(1 - |S_{11}|^2\right)\left(1 - |S_{22}|^2\right)} \tag{3.6}$$

where $i = 1, 2, 3 \dots 6$.

An average level can then be formed from the x orientations according to (3.7).

$$P_{AV\_x} = \frac{1}{6}\sum P_{i,x} \tag{3.7}$$

Similarly, for the y and z orientations, similar averages can be formed as (3.8) and (3.9) by using y and z orientated measurement data in the form of (3.6).

$$P_{AV\_y} = \frac{1}{6}\sum P_{i,y} \tag{3.8}$$

$$P_{AV\_z} = \frac{1}{6}\sum P_{i,z} \tag{3.9}$$

A total polarisation reference level can then be defined as [11]:

$$P_{POL\_REF} = \frac{1}{3}\sum P_{AV\_x,y,z} \tag{3.10}$$

The polarisation imbalance for the x, y and z orientations can then be expressed as (3.11).

$$\frac{P_{AV\_x}}{P_{POL\_REF}}, \frac{P_{AV\_y}}{P_{POL\_REF}}, \frac{P_{AV\_z}}{P_{POL\_REF}} \tag{3.11}$$

Figures 3.23 and 3.24 depict the measured polarisation imbalance in decibel scale as a function of frequency between the standard paddles and the new paddle design.

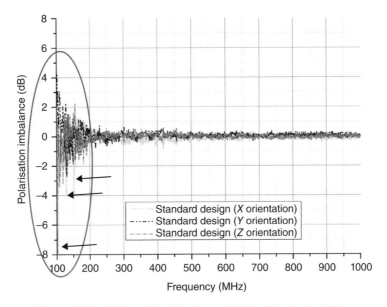

Figure 3.23   Polarisation imbalance (dB) for standard paddles.

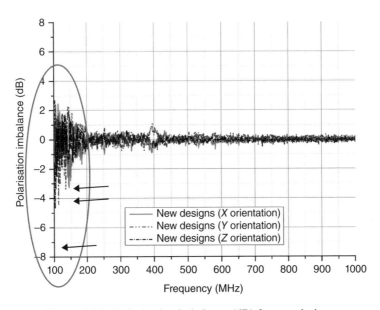

Figure 3.24   Polarisation imbalance (dB) for new designs.

A polarisation imbalance is stated to result from both unstirred direct power and a modal excitation error in the chamber [11]. The latter is stated to appear as a result of the chamber and the mechanical stirrers being too regularly shaped, causing the TE and TM modes not to mix when stirred [11]. Hence, it is important to minimise this effect such that the amount of received power in the chamber does not depend upon an antenna's orientation.

Comparing Figures 3.23 and 3.24 it can be seen that there is an improvement in performance from the new designs in the 100–200 MHz range, being signified by the arrows. In the z orientation there is a 4 dB improvement witnessed in some cases, and 1–1.5 dB improvements are witnessed for the x and y orientations. In the lower frequency regions (<200 MHz), a relatively high imbalance was expected due to the sparseness in the overall number of modes available as depicted previously in Chapter 2.

However, it can be seen that due to the extreme irregular nature of the cuts that are configured to be resonant in this lower frequency region, in particular in the two principle planes, has served to minimise the polarisation discrepancy. At higher frequencies it can be seen that both standard and new designs perform perfectly – virtually no imbalance is present.

The strange peak that is evident within the new designs at 400 MHz is believed to be due to an anomaly in the resonant performance of the cuts from the new paddle sets at this frequency – it is also present where the standard deviations are concerned (Figure 3.22). One way to eliminate this effect would be to configure a specific length cut at this frequency. If adopting such a technique, a designer should be aware of any such anomalies and configure the electrical performance of the paddles accordingly.

The Rician K factor performance of the standard and new designs is detailed in Figure 3.25. Please refer Chapter 2 for the derivation of this quantity.

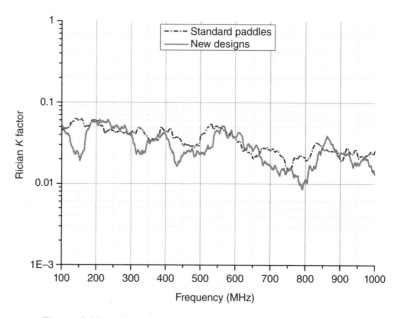

Figure 3.25   Rician K factor comparison in unloaded chamber.

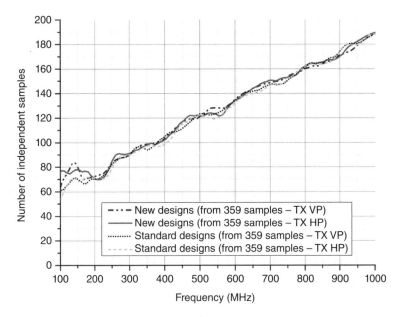

Figure 3.26   Number of independent samples comparison in unloaded chamber.

Please note that a separate measurement sequence has been undertaken for the deduction of this and the following quantity. In this, 1 degree mechanical stirring, polarisation stirring and 1× receiver location has been applied. The antenna has also been orientated in the z direction. All other parameters in Table 3.7 apply.

From Figure 3.25 it can be seen that generally there is a lower proportion of direct power being registered at the receiver. Particularly, at lower measured frequencies, there is a marked difference. This is believed to be attributed to the added resonant capability of the new stirrer designs, in the fact that there are specific length cuts on the paddle that promote this resonant behaviour. Further evidence to support this statement can be found in Appendix C where the current distributions on the stirrer paddles are analysed.

For the NLoS number of independent samples, the full procedure is detailed in Appendix A. Using the 1 degree mechanical stirred, polarisation stirred and 1× receiver location acquired data; Figure 3.26 details the comparison between the standard and new designs for each measured transmitting polarisation.

From Figure 3.26 it can be seen that in the 100–200 MHz range there is an improvement in the overall number of samples that can be considered independent (sufficiently uncorrelated) with the new designs. This improvement is seen to be registered for both transmitting polarisations which would suggest that the cuts on the paddle in orthogonal planes have had a positive effect in stirring both polarised waves.

Beyond 200 MHz, in a region where the original paddle dimensions can generally be considered to be electrically large, performance is seen to be more or less comparable, although in areas the new designs appear to be better performing.

To conclude this subsection, four different performance benchmarks have been applied to assess the performance merits of the new designs as compared to a standard design of paddle. The four separate benchmarks have all confirmed improved performance for the new designs in the lower modal region which serves to provide confidence that the design strategy does indeed work and does improve the chamber performance. For all measurements down at this lower modal region, quantities could be acquired with less uncertainty; and where the standard deviations are concerned, using less measurement samples (and less time) to yield the same overall performance levels; these represent the benefits of undertaking such a modification.

The final subsection of this chapter repeats all the aforementioned results but in a loaded chamber. The purpose of assessing this is because during some measurements (e.g. large array antennas or large pieces of equipment undergoing EMC tests), the chamber can be presented with significant loading which will lower its Q factor. The chamber should always be characterised in both scenarios to provide a complete picture of its true performance across a range of applications and measurement scenarios.

3.6.2 Standard vs New Designs: Loaded Chamber

In this section all the aforementioned measurement parameters and details still apply. The only factor that has changed is that the chamber is loaded with four large pieces of anechoic absorber, two each in the back two corners of the chamber to simulate the 'loading'.

The standard deviation comparison in the loaded scenario is detailed in Figure 3.27.

From Figure 3.27 it is immediately apparent the effect that loading has had on chamber performance; the standard deviations in this case are typically 0.25 dB higher. Towards higher frequencies this difference becomes slightly lower – typically in the order of 0.15 dB. The subsequent reason for the increase in the standard deviation with the loading is because the pieces of anechoic absorber can serve to dampen given resonant modes depending on where in the chamber that loading is situated [9]. If a given resonant mode is damped, this will have a consequence on the nature of the field distribution in the chamber as some of the electric field peaks can be suppressed. At given stirrer increments therefore, this can cause the field value not to vary significantly which in turn can result in a larger deviation inherent from different receiver antenna locations as seen.

In this loaded scenario it can be seen that the new paddle still outperforms the standard design. At higher frequencies this time the standard deviation for the new

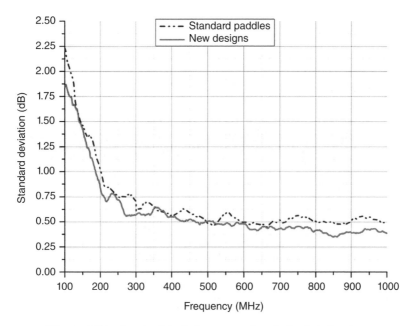

Figure 3.27   Standard deviation comparison in loaded chamber.

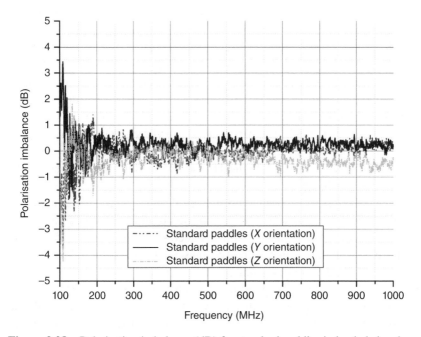

Figure 3.28   Polarisation imbalance (dB) for standard paddles in loaded chamber.

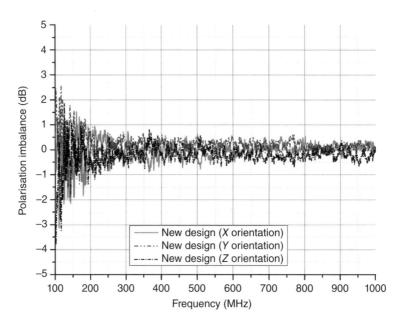

Figure 3.29   Polarisation imbalance (dB) for new designs in loaded chamber.

paddles shows better performance than the standard design which is believed to be due to the nature of the resonant cuts in two principle planes that has managed to stir both the TE and TM modes more efficiently about the larger average mode bandwidth.

The polarisation imbalance can be displayed as in Figures 3.28 and 3.29. On the whole, when comparing both figures it can be seen that the new stirrer designs again outperform the standard design. From Figure 3.28 it can be seen that a polarisation imbalance for the standard design in the z orientation is prevalent all the way out to 1000 MHz – this coincides with the larger standard deviations inherent from Figure 3.27. Again, this is believed to be due to the nature of the resonant cuts in two principle planes that has managed to stir both the TE and TM modes more efficiently about the larger average mode bandwidth.

The Rician K factors for both standard and new designs can be viewed in Figure 3.30, from which it can be seen that first, the proportion of direct power has increased with the additional loading as compared to the unloaded scenario (Figure 3.25). Irrespective of this, it can be seen that the new stirrer designs again yield better performance on the whole than a standard design of paddle.

The NLoS number of independent samples can be viewed in Figure 3.31.

From Figure 3.31 it can be seen that at lower frequencies an increase in the number of independent samples is evident, similar to the unloaded investigation. Overall, it can be seen that the total number of samples that are considered to be independent is

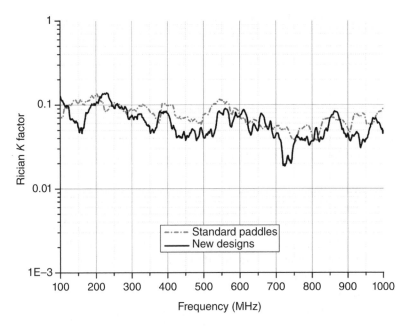

Figure 3.30   Rician K factor comparison in loaded chamber.

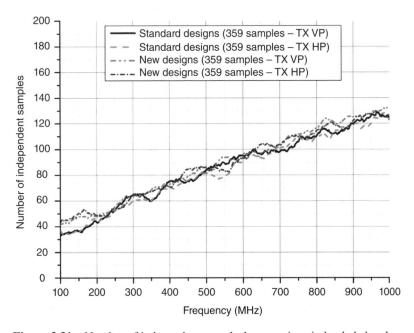

Figure 3.31   Number of independent samples' comparison in loaded chamber.

lower than for the unloaded case, proving that the mechanical stirring process finds it more difficult to significantly alter the field distribution about an enlarged average mode bandwidth.

3.7 Summary

In this chapter, a new design of mechanical stirring paddle has been theoretically and numerically developed. Practically, a rigorous assessment has been undertaken to compare the merits of the new design against a standard design of paddle that is common to most RCs.

The theoretical idea behind the design has advocated a series of cuts on the paddle structure with a view to lower the resonant capability of the paddle structure by increasing the current path length. In addition, many of these cuts were tried and tested, each with varying lengths in an attempt to enact a broadband enhanced performance. An alternative numerical approach was adopted that yielded an improved design in an efficient manner.

Practical results showed that the theoretical idea worked and performance was improved. The performance aspect was validated by four separate performance indicators whose assessment all came back in favour of the new designs. Given the nature of the cuts on the modified paddle, the structural rigidity of the paddles should be carefully considered to prevent excessive flexing after movement.

With regards to the chamber performance validation in practice, we have detailed the procedures and equations required to fully characterise the chamber performance. The procedures detailed will provide more of a robust representation of the true performance of a chamber, as multiple performance parameters are used and compared. The procedures should also apply to the EMC community for RC measurement uncertainty analysis but for completeness, the conventional (standards) chamber validation procedure is presented in Appendix D.

At this time, the consensus in thinking is that the new procedures could add to the chamber validation protocol in the BS EN 61000-4-21 standards. This raises interesting questions which will ultimately be down to the community to decide.

References

[1] Y. Huang, J. T. Zhang and P. Liu, 'A novel method to examine the effectiveness of a stirrer', *2005 International Symposium on Electromagnetic Compatibility, 2005*. EMC 2005, vol. 2, IEEE, 12 August 2005, Chicago, IL, pp. 556–561.
[2] Y. Huang, N. Abumustafa, Q. G. Wang and X. Zhu, 'Comparison of two stirrer designs for a new reverberation chamber', *The 2006 4th Asia-Pacific Conference on Environmental Electromagnetics*, IEEE, 1–4 August 2006, Dalian, pp. 450–453.

[3] J. I. Hong and C. S. Huh, 'Optimization of stirrer with various parameters in reverberation chamber', *Progress In Electromagnetics Research*, vol. 104, pp. 15–30, 2010.

[4] N. Wellander, O. Lunden and M. Backstrom, 'Experimental investigation and mathematical modeling of design parameters for efficient stirrers in mode-stirred reverberation chambers', *IEEE Transactions on Electromagnetic Compatibility*, vol. 49, pp. 94–103, 2007.

[5] L. R. Arnaut, 'Effect of size, orientation, and eccentricity of mode stirrers on their performance in reverberation chambers', *IEEE Transactions on Electromagnetic Compatibility*, vol. 48, pp. 600–602, 2006.

[6] J. Clegg, A. C. Marvin, J. F. Dawson and S. J. Porter, 'Optimization of stirrer designs in a reverberation chamber', *IEEE Transactions on Electromagnetic Compatibility*, vol. 47, pp. 824–832, 2005.

[7] D. I. Wu and D. C. Chang, 'The effect of an electrically large stirrer in a mode-stirred chamber', *IEEE Transactions on Electromagnetic Compatibility*, vol. 31, pp. 164–169, 1989.

[8] S. R. Best, 'On the resonant properties of the Koch fractal and other wire monopole antennas', *IEEE Antennas and Wireless Propagation Letters*, vol. 1, pp. 74–76, 2002.

[9] Y. Huang, *'The Investigation of Chambers for Electromagnetic Systems'*, D Phil Thesis, Department of Engineering Science, University of Oxford, 1993.

[10] CST Microwave Studio, http://www.cst.com (accessed 28 July 2015).

[11] BS EN 61000-4-21:2011 In Electromagnetic compatibility (EMC) Part 4-21: Testing and measurement techniques — Reverberation chamber test methods, ed: BSI Standards Publication, 2011.

[12] P. S. Kildal, X. Chen, C. Orlenius, M. Franzen and C. S. L. Patane, 'Characterization of reverberation chambers for OTA, measurements of wireless devices: Physical formulations of channel matrix and new uncertainty formula', *IEEE Transactions on Antennas and Propagation*, vol. 60, pp. 3875–3891, 2012.

[13] P. S. Kildal, C. Orlenius, J. Carlsson, U. Carlberg, K. Karlsson and M. Franzen, 'Designing reverberation chambers for measurements of small antennas and wireless terminals: Accuracy, frequency resolution, lowest frequency of operation, loading and shielding of chamber', *First European Conference on Antennas and Propagation. EuCAP 2006*, IEEE, 6–10 November 2006, Nice, pp. 1–6.

[14] A. Coates and A. P. Duffy, 'Maximum working volume and minimum working frequency trade-off in a reverberation chamber', *IEEE Transactions on Electromagnetic Compatibility*, vol. 49, pp. 719–722, 2007.

[15] L. R. Arnaut, 'Operation of electromagnetic reverberation chambers with wave diffractors at relatively low frequencies', *IEEE Transactions on Electromagnetic Compatibility*, vol. 43, pp. 637–653, 2001.

[16] J. R. Taylor, *An Introduction to Error Analysis: The Study of Uncertainties in Physical Measurements*, 2nd ed.: Sausalito: University Science Books, 1997.

4

EMC Measurements Inside Reverberation Chambers

In the previous two chapters, we have discussed the Reverberation Chamber (RC) cavity theory, stirrer design and the RC characterisation. We will continue our discussion on the RC by introducing some important chamber parameters related to its operation, such as the Lowest Usable Frequency (LUF) and the working volume, which is also known as the Equipment Under Test (EUT) area, but the focus will be placed on Electromagnetic Compatibility (EMC) measurements and tests inside RCs – this was the very reason why the RC was originally introduced to the EMC community in 1968 [1, 2].

In this chapter, the basics of EMC will be introduced first which will include some essential concepts, definitions, fundamentals and activities of EMC. It will then be followed by the introduction of EMC standards and measurements. The radiated emission and immunity/susceptibility tests will be discussed in detail where the IEC standard, IEC61000-4-21 [3], will be used as the main reference – since it contains the most widely accepted EMC testing and measurement techniques inside an RC. Measurement procedures and error analyses will be presented. A comparison of EMC measurements inside an RC and other conventional facilities will be made at the end of this chapter.

The aim of the chapter is to link the electromagnetic and RC theories to practical EMC measurements and tests. Some examples and references are used to serve this objective.

Reverberation Chambers: Theory and Applications to EMC and Antenna Measurements, First Edition.
Stephen J. Boyes and Yi Huang.
© 2016 John Wiley & Sons, Ltd. Published 2016 by John Wiley & Sons, Ltd.

4.1   Introduction to EMC

Electromagnetic Interference (EMI) has been a well-known phenomenon for a very long time, while Electromagnetic Compatibility (EMC) is a relative new term for many people. The IEEE standard definitions [4] on EMI and EMC are as follows:

> **EMI** is the degradation of the performance of a transmission channel, equipment, or system caused by an electromagnetic disturbance.
>
> **EMC** is the ability of a device, equipment or system to function satisfactory in its environment without introducing intolerable electromagnetic disturbances to anything in that environment.

This means that

- The product does not cause interference with other systems (so other systems can work satisfactory).
- It is not susceptible to the emission from other systems (so it works satisfactory in its environment).
- It does not cause interference with itself (so it works satisfactory).

It is apparent that EMC has taken EMI into account and is a subject of considering Electromagnetic (EM) emission and susceptibility (or immunity) of a product. The EM *emission* is about unwanted EM signals emitted or transmitted through a conducted path (such as a cable) or a radiated path (a radio wave) over the frequencies of interest. The EM *susceptibility* of a product is about how sensitive/susceptive it is to incoming EM signals via a conducted or a radiated path. Similarly, the EM *immunity* of a product is about how insensitive (not susceptive) it is to incoming EM signals via a conducted or a radiated path. Thus a good product should have low susceptibility but high immunity to EMI. Since the susceptibility and immunity are complementary concepts and deal with the same issue: to ensure that a product can work well in a harsh EM environment, we therefore consider susceptibility and immunity as the same aspect of EMC. Because the EMI signals can affect a product via a conducted and/or a radiated path, thus the EMC activities are normally divided into conducted and radiated parts as shown in Figure 4.1. In the conducted part, EMC issues are caused via conducted paths (such as cables), whereas in the radiated part, EMC problems are caused via radiated paths (such as radio waves). For each part, one should deal with both the emission and immunity (susceptibility) issues, thus any EMC problem could fall into one or more of the following four aspects:

1. Conducted emission
2. Conducted immunity/susceptibility
3. Radiated emission
4. Radiated immunity/susceptibility

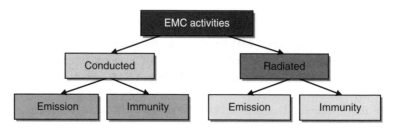

Figure 4.1 The main EMC activities.

A conducted emission and/or immunity problem occurs normally at lower frequencies, whereas a radiated emission and/or immunity problem happens at higher frequencies. This is because at lower frequencies (typically <30 or 80 MHz):

- Conducted signals (voltages and currents) can travel over a long distance without much attenuation via a conducted path;
- Conducted signals are relatively easy to measure;
- Radiated signals are normally very small (the product sizes are normally very small compared with the wavelength) and also very hard to measure.

But at higher frequencies (typically >30 or 80 MHz):

- Conducted signals (voltages and currents) cannot travel over a long distance due to the attenuation and radiation via a conducted path;
- Radiated signals could be relatively large (the product sizes might be comparable with the wavelength) and easy to measure.

Thus in practice, at lower frequencies, the conducted emission and immunity are of major concern while at higher frequencies, the radiated emission and immunity are considered. It should be pointed out that both the conducted and radiated problems may occur at any frequency. These classifications are well justified and suitable for practical implementation. At lower frequencies, if there is no conducted emission or immunity problem, there should be no radiated emission or immunity problem. Similarly, at higher frequencies, if there is no radiated problem, there should be no conducted emission or immunity problem.

By examining EMC problems, one can find that any EMC problem consists of three elements: source, transmission and reception of EM energy. They form the basic framework of EMC analyses and designs. The basic decomposition of an EMC problem is illustrated in Figure 4.2. The source (also referred to as an emitter which can be a man-made or a natural source, such as lightning) of EMI produces the emission of EM energy, and a conducted or radiated coupling path transfers the emission energy to

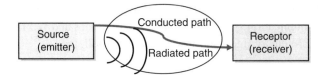

Figure 4.2   The basic components of an EMC coupling problem.

a receptor (receiver), where it is processed, resulting in either a desired or undesired behaviour. Interference occurs if the received energy causes the receptor to behave in an undesired manner. Transfer of EM energy occurs frequently via unintended coupling modes. However, the unintentional transfer of EM energy causes interference only if the received energy is of sufficient magnitude and/or spectral content at the receptor input to cause the interference or to behave in an undesired fashion. Unintentional transmission and reception of EM energy are not necessary detrimental. That is why EMC is not just about how to deal with the emission (EMI), but also about how to deal with the susceptibility and immunity. The discussion here suggests that there are three general ways to solve an EMC problem:

1. suppress emission at its source
2. make the coupling path as inefficient as possible
3. make the receptor less susceptible (better immunity) to the emission

It should be pointed out that EMC has actually been around for almost 100 years but not everyone has realised it. Technical papers on radio interference began to appear in various journals around 1920. During the Second World War, the use of electronic devices, primarily radios, navigation devices and radar accelerated. Instances of interference between radios and navigational devices on aircraft began to increase. These were easily corrected by reassignment of transmitting frequencies of un-crowded spectrum or physically moving cables away from the noise emission source to prevent the cables from picking up those emissions. These interference remedies can be easily implemented on a case-by-case basis in order to correct an EMI problem. However the most significant increases in the interference problem occurred with the inventions of high-density electronic components, such as the bipolar transistor in the 1950s, the integrated circuit in the 1960s and microprocessor chip in the 1970s. The frequency spectrum also became more crowded due to the increased demand for voice and data transmission, wireless communication systems and wireless sensor networks. This required considerable planning with regard to spectrum utilisation and it continues to be so even today.

Perhaps the primary event that brought the present emphasis on EMC to the forefront was the introduction of high-speed digital communications, processing and computation. The density and bandwidth of noise sources in spectral content became quite large.

Consequently, the occurrence of EMI problems began to rise. Nowadays the widespread use of radio communication systems such as cellular phones and WiFi systems has certainly made the EM environment more congested and polluted.

4.2 EMC Standards

To clean up the EM environment and set limits on EM emissions, a meeting of International Electromechanical Commission (IEC) in Paris recommended the formation of the International Special Committee on Radio Interference (CISPR) to deal with the emerging problem of EMI in 1933. Subsequent meetings yielded various technical publications, which dealt measurement techniques as well as recommended emission limits and immunity levels. A wide range of standards and regulations have been introduced over the years around the world to deal with EMC problems. One of the most significant events was the introduction of the European Commission (EC) Directive on EMC in 1989 [5], which was fully implemented in 1996. It basically means that all electrical and electronic products sold in the European Union (EU) must meet the EMC requirements set by the EU. This Directive on EMC has made the whole industry (and the world) aware of the importance of EMC and served as a major driving force to improve product EMC performance. Over the years, many EMC standards have been introduced by different standards- making bodies. The most influential one is the IEC, which has produced comprehensive standards on EMC. Most of them are transposed to national or harmonised European Norm (EN) standards. For example, IEC61000 is a set of basic EMC standards as shown in Table 4.1 and has been transposed to European standard as EN61000 and to British standard as BS EN 61000. It has set emission and immunity limits as well as provided detailed information on how to conduct relevant tests and measurements. Generally speaking, EMC standards can be divided into generic standards (such as EN50081 for emission and EN50082 for immunity), basic standards (e.g. IEC 61000) and product standards (e.g. EN 55014 for household appliances, electric tools and similar apparatus, EN 55022 and EN 55024 for information technology equipment).

Table 4.1 Basic Standard IEC 61000.

Standard	Description
IEC 61000-1	Part 1: General
IEC 61000-2	Part 2: Environment
IEC61000-3	Part 3: Limits (including emission limits, immunity limits)
IEC 61000-4	Part 4: Testing and measurement techniques
IEC 61000-5	Part 5: Installation and mitigation guidelines
IEC 61000-6	Part 6: Generic standard
IEC 61000-9	Part 9: Miscellaneous

Different products may need to apply different standards – this is one of the complications of implementing EMC standards and regulations in practice.

In addition to these national and international standards, there are also military standards for military platforms (such as ships, aircraft and vehicles), weapons, equipment and systems. For example, in the United States, MIL-STD-461D has specified a variety of levels and limits for different purposes, and MIL-STD-462D has defined the corresponding test methods. These two were merged into MIL-STD-461E in 1999 (with a number of changes). In the United Kingdom, the DEF STAN 59-41 series provides a similar variety of tests to the US documents but in a different format, and also gives project planning and documentation requirements along with installation guidelines. All military standards tend to have more stringent requirements than civil standards over a wider frequency range due to the possible harsh environment and close proximity in a platform.

EMC standards have played a very important role in our daily life by ensuring good EM environments and reliable quality products. The conducted emission limits and radiated emission limits set by EN 550XX series (also CISPR 22) and American FCC (Federal Communications Commission) are shown in Figures 4.3 and 4.4, respectively. They are also given in Tables 4.2 and 4.3, respectively. *Class A* is for use in all establishments (such as industrial) other than domestic and *Class B* is suitable for use in domestic establishments. The Class B limits are more stringent than the Class A limits under the reasonable assumption that interference from the device under industrial environment can be corrected more easily than in a domestic environment, where the interference source and the

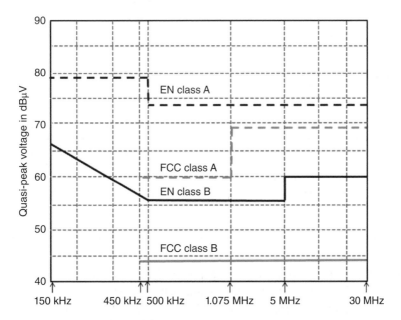

Figure 4.3 Conducted emission limits.

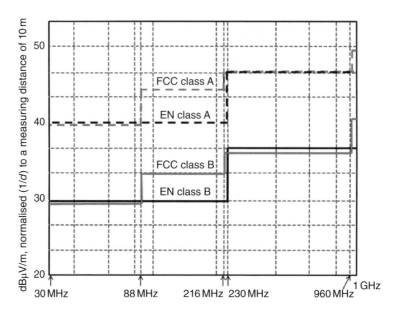

Figure 4.4   Radiated emission limits.

Table 4.2   Conducted emission limits.

Class A	FCC limits	EN/CISPR limits	EN/CISPR limits
Frequency (MHz)	Voltage quasi-peak	Voltage quasi-peak	Average (dBμV)
0.15–0.45	No limits	79	66
0.45–0.5	60	79	66
0.5–1.705	60	73	60
1.705–30	69.5	73	60
Class B	FCC limits	EN/CISPR limits	EN/CISPR limits
Frequency (MHz)	Voltage quasi-peak	Voltage quasi-peak	Average (dBμV)
0.15–0.45	No limits	66–56.9	56–46.9
0.45–0.5	48	56.9–56	46.9–46
0.5–5	48	56	46
5–30	48	60	50

susceptible device are likely to be in closer proximity. FCC levels differ somewhat from the harmonised EN (and CISPR) levels and are included for comparison – this also highlights the feature of the current localised EMC standards (there is a need to establish global standards, although this is difficult). All radiated emissions are normalised to a

Table 4.3 Radiated emission limits.

Class A	FCC limits	EN/CISPR limits	
Frequency (MHz)	Field strength (dBμV/m) at 10 m	Field strength (dBμV/m) at 10 m	
30–88	39	40	
88–216	43.5	40	
216–230	46.4	40	
230–960	46.4	47	
960–1000	49.5	47	
>1000	49.5	No limit	

Class B	FCC limits	EN/CISPR limits	
Frequency (MHz)	Field strength (dBμV/m) at 3 m	Field strength (dBμV/m) at 10 m	Field strength (dBμV/m) at 10 m
30–88	40	29.5	30
88–216	43.5	33	30
216–230	46	35.6	30
230–960	46	35.6	37
960–1000	54	43.5	37
> 1000	54	43.5	No limit

measuring distance of 10 m – the field specified at other distances, such as 3 m, can be easily converted to 10 m since the field is inversely proportional to the distance. There are detailed differences in the measurement methods between EN/CISPR and FCC standards. The conducted frequency range is 150 KHz to 30/80 MHz for EN but 450 KHz to 30/80 MHz for FCC. For immunity tests, the levels could be very different. For example, EN 61000-4-3 has set severity levels of 1, 3 and 10 V/m from 30/80 to 1000 MHz. The frequency boundary between the conducted and radiated emission limits was 30 MHz but is now 80 MHz in most standards.

Of course, EMC standards are not just about setting the conducted and radiated emission limits and immunity levels. As shown in Table 4.1, they cover a wide range of activities and scenarios. But the emission measurements and immunity tests are the most critical parts of the enforcement of EMC standards.

4.3 EMC Measurements and Tests

There are a wide range of EMC measurements and tests. Generally they can be divided into four aspects: conducted emission measurements, conducted immunity tests, radiated emission measurements and radiated immunity tests. For most conducted emission measurements and immunity tests, the required facilities and equipment are relatively

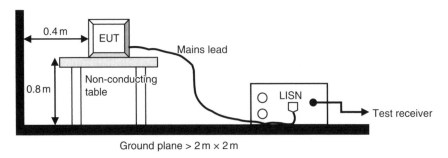

Figure 4.5   A typical layout for conducted emission measurements.

simple. For example, the only vital requirement for most conducted measurements and tests is a conducting ground plane of at least $2\,m \times 2\,m$ by many standards, extending at least $0.5\,m$ beyond the boundary of the EUT. It is recommended, but not essential, to make the measurement in a conducting screened room to reduce the background noise. The equipment required for conducted emission measurements are a line impedance stabilisation network (LISN, relatively cheap special equipment) and a spectrum analyser or receiver. A typical layout for conducted emission measurements is shown in Figure 4.5. For conducted immunity tests, an LISN is not used but a bulk current is injected or coupled into the EUT, a test method can be found from a basic standard such as IEC 61000-4-6.

In contrast with the conducted measurements and tests, the radiated measurements and tests are much more complicated and time-consuming. They are normally expected to be undertaken in either an anechoic chamber (AC) or a semi-anechoic chamber. A typical AC is a conducting screened room lined with radio absorbing materials (RAMs) on the walls, ceiling and floor (thus there are no reflections or echoes inside the chamber) to simulate the free space case, while a semi-anechoic chamber, as shown in Figure 4.6, is the same as an AC except that there are no RAMs on the floor to simulate the open space with a ground floor case. The major advantages of using these chambers are

1. indoor environments not affected by the weather
2. isolation from the outside, thus they do not produce interference to the outside environment and also are not interfered by external sources and
3. very accurate and repeatable measured results in different chambers

The EUT should be placed in a 'quiet zone' where the EM wave can be considered as a plane wave (the field is uniform). Figure 4.6 shows a typical layout for radiated emission measurements where the measurement distance L is defined by the relevant standard of the test (the typical values are 3 and $10\,m$), while the antenna height H should be varied from 1 to $4\,m$ and the EUT should be rotated at each test frequency for different orientations/polarisations. Similarly, for the radiated immunity tests, varied antenna positions and EUT orientations/polarisations should be tested at each

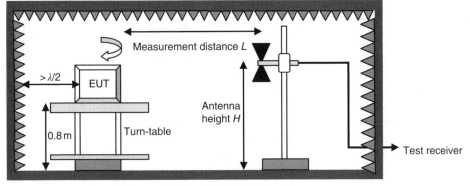

Figure 4.6 A typical layout for radiated emission measurements.

frequency, and expensive broadband power amplifiers are needed to generate the high field levels inside the chamber. Since the frequency range of interest is normally very broad, typically between 80 and 1000 MHz, hence the radiated emission and immunity measurements and tests can be a very long and expensive process.

In addition to ACs and semi-anechoic chambers, other test facilities have also been proposed and used for radiated EMC measurements, such as the TEM-cell [6] and GTEM-cell [7], which are good alternatives for small EUTs. Another good alternative facility is an RC.

4.4 EMC Measurements Inside Reverberation Chambers

As a conducting screened room, an RC can be used for making conducted EMC measurements and tests without the need to stir the field inside the chamber. The main question and focus of this part is if an RC can be used for radiated EMC measurements and tests. If yes, how to do it? And how to interpret the results?

Unlike an AC, an RC is an electrically large conducting chamber with at least one mechanical stirrer to stir or tune the EM field inside the chamber. It has reflective boundary conditions, and radio waves are reflected (not absorbed as an AC) by the walls, floor and ceiling, thus various resonant modes and field patterns are generated which did not seem to be useful for EMC measurements. Over the years, many research groups have studied the RC and found that it is possible to perform both radiated emission and radiated immunity EMC measurements inside an RC. The field has changed significantly from one point to another point. However, due to the movement/rotation of the stirrer, the averaged field in the EUT area is relatively uniform (homogenous and isotropic). The polarisation of this field varies randomly which results in that the avoidance of the physical rotation of the EUT – this means a reduction of the measurement time and also an easier approach especially to measure a large EUT. Another very attractive feature is that it is capable of providing a very efficient conversion of the supplied power to high-intensity fields inside the chamber without the need for expensive broadband high-power amplifiers (as for an AC) – this is due to the conductive boundary feature of the chamber: the loss of the chamber is normally small, thus it has a high quality factor (Q-factor). Overall, it has shown that an RC can offer at least the following advantages when used for EMC measurements:

1. Excellent electrical isolation from the external environment
2. Excellent accessibility (indoor test facility)
3. Very broad frequency range (the operational frequency should be greater than the lowest usable frequency of the chamber)
4. Easy to generate high-intensity fields efficiently
5. No requirement on physical rotation of the EUT (due to the randomness of the field polarisation inside the chamber)
6. Cost effectiveness

But it also has some limitations, which includes the difficult interpretation of the measurement result, loss of the polarisation and direction information on the EMC profile of the EUT. Nevertheless, this facility offers an attractive alternative technique for radiated EMC measurements and tests from both the technical and economical points of views.

4.4.1 Relevant EMC Standards Using Reverberation Chambers

Since the investments in an AC infrastructure are high and the EMC measurements to be performed are complex and time-consuming, the use of RC for EMC measurements is therefore a very attractive alternative. It has now been adopted by some standards. A basic standard, IEC 61000-4-21 [3], is the most widely available and comprehensive document about the RC validation, radiated immunity tests, and radiated emission measurements. A recommended typical RC for EMC measurements is shown in Figure 4.7.

There are also manufacturer-specific standards using RCs for EMC measurements. For example, several SAE (Society of Automotive Engineers) standards cover RC radiated immunity testing on vehicles and components:

1. J551-16: Vehicle-level radiated immunity tests (mode-tuned and mode-stirred methods which will be discussed in next section)
2. J1113-27: Component-level radiated immunity tests (mode-stirred method)
3. J1113-28: Component-level radiated immunity tests (mode-tuned method)

In addition, GMW 3097 and Ford ES-XW7T-1A278-AC are also for the automotive EMC measurements, and RTCA DO 160D is for aerospace (a standard about 'Environmental Conditions and Test Procedures for Airborne Equipment' published by RTCA in 1997) [8].

RCs have also been approved for military EMC measurements in accordance with MIL-STD-461E.

At the moment, the acceptance of using RCs for EMC measurements is still not very high. As the technology develops further, a wider acceptance by various industries is expected.

4.4.2   Chamber Characterisation

Chapter 2 has introduced the RC theory, some major chamber characteristics and parameters, such as the quality Q factor. In this section, some more parameters and characteristics of an RC is introduced and discussed since they are closely linked to EMC measurements and tests.

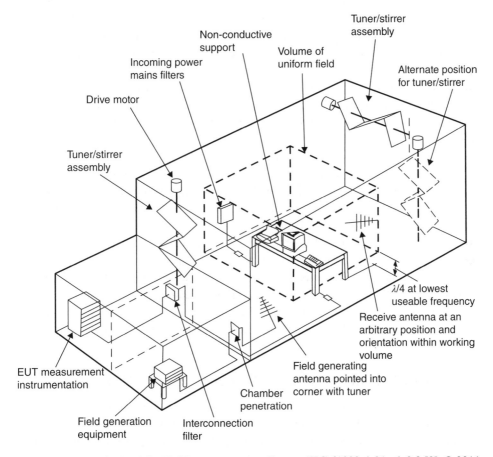

Figure 4.7   A typical RC for EMC measurements. Source: IEC 61000-4-21 ed. 2.0 [3]. © 2011 IEC Geneva, Switzerland.

4.4.2.1   Mode-stirred and Mode-tuned Operations

The stirrer/tuner inside an RC can be moved continuously or one-step a time when performing EMC testing. The first method is called 'mode-stirred' (also known as continuous-mode) operation where the field inside the chamber is changing while the measurement is taking place. The second one is called 'mode-tuned' (also known as step-mode) operation where the field is stable when the measurement or sampling is taking place.

In mode-tuned operation, the stirrer is fixed at particular rotational angles while the measurements are made – this ensures that the EUT is exposed to an unchanging field for any desired length of time. However, physically stopping the stirrer for each measurement increases the overall measurement time. Moreover, the starting and stopping of the stirrer lead to mechanical vibrations that can introduce a crackling effect [9] that produces large transients in the field. Measurements must be delayed until these mechanically induced transients die down. Due to the large measurement time associated with stepping, the standards allow as few as 12 unique stirrer positions (as we will see later) that is used in qualification testing which results in the industry qualifying their products with a test that exposes the EUT for a limited number of aspect angles and polarizations leading to the increase in the variance of the measurements [10].

In mode-stirred operation, the stirrer is continuously rotated while the measurements are performed with no stops at specific angles. This significantly reduces the testing time compared to stepping and therefore is cost effective. Moreover, the crackling effect is eliminated since there is no mechanical backlash associated with the starting and stopping of the stirrer, and there is no acceleration/deceleration once the stirrer reaches its desired rotational speed. Finally, stirring exposes the EUT to the maximum number of field configurations possible within a given chamber at the operating frequency when the stirrer is rotated through a full turn. However, a concern with mechanical stirring is that the EUT may not be exposed to the same/required field level for a sufficient length of time to provide full susceptibility testing.

Both operations have their pros and cons and are accepted by standards such as IEC 61000-4-21 for radiated emission and immunity measurements. Since most research work conducted in this area is based on the mode-tuned operation, our discussion in this chapter will be mainly based on this operation unless otherwise stated.

4.4.2.2   Field Uniformity, Working Volume and Lowest Usable Frequency

The goal of an RC is to generate a statistically uniform environment for all locations within the defined working volume, also known as the EUT area. But how to define this working volume could be a problem since it is a function of the frequency and also a function of the definition or the requirements of the field uniformity. These are some of the most important questions we have to answer before undertaking any EMC measurements and tests.

In Figure 4.7, a rectangular EUT area is shown which should be away from the floor, walls, ceiling and other objects (such as the stirrer) inside the chamber by a ¼ wavelength of the lowest usable frequency (LUF). In IEC 61000-4-21, the uniformity of this working volume is determined by the fields collected using an electric field (E-field) probe at eight corner locations of the rectangular volume. At each location, three orthogonal E-fields are sampled over the frequency band of interest for every stirrer step of a revolution. The total data of the E-fields could be huge and the process could be excessive. To make this calibration process more manageable, the IEC 61000-4-21:2011 [3] has set the minimum sampling requirements as shown in Table 4.4. f_s is the starting frequency which basically means the LUF of the chamber. The whole frequency range is divided into four frequency bands and each frequency band should have at least 10–20 frequency samples. At each sampling frequency, the minimum number of stirrer steps is 12 for all frequencies. For many chambers, the number of stirrer steps will need to be increased at the lower frequencies. For example, the recommended number of stirrer steps in the first edition of the IEC 61000-4-21:2003 (the same as EN61000-4-21:2003 [11]) is larger at lower frequencies (from f_s to $6f_s$) and is also shown in the table, in bracket, for comparison. Increasing the number of stirrer steps increases the expected value of the maximum field and reduces the uncertainty in the data but it takes a longer time [3, 11].

To define the field uniformity, the field standard deviations at all eight corner locations are employed and the tolerance requirements for standard deviation over the frequency range of interest are set in Table 4.5. The whole frequency range has been divided into three bands. The tolerance is 4 dB for the lower band (80–100 MHz) and 3 dB for the higher band (above 400 MHz). The tolerance between these two bands is decreased linearly from 4 to 3 dB.

An example of the standard deviations of measured data for the three E-field components (x, y and z) and the total E-field from eight corner locations of the working volume is shown in Figure 4.8. We can see that the standard deviation is well within the tolerance, above 200 MHz, for all components. But it is over the limit when the frequency is down to about 120 MHz, thus the LUF for this case is 120 MHz. Here we

Table 4.4 Sampling requirements (f_s is the starting frequency, or LUF).

Frequency range	Min number of stirrer steps required (recommended)	Number of frequencies required
f_s to $3f_s$	12 (50)	20
$3f_s$ to $6f_s$	12 (18)	15
$6f_s$ to $10f_s$	12 (12)	10
Above $10f_s$	12 (12)	20/decade

The numbers in parenthesis are recommended from IEC 61000-4-21:2003 and all the others are from IEC 61000-4-21:2011.

Table 4.5   Field uniformity tolerance requirements.

Frequency range MHz	Tolerance requirements for standard deviation
80–100	4 dB
100–400	4 dB at 100 MHz decreasing linearly to 3 dB at 400 MHz
Above 400	3 dB

A maximum of three frequencies per octave may exceed the allowed standard deviation by an amount not to exceed 1 dB of the required tolerance.

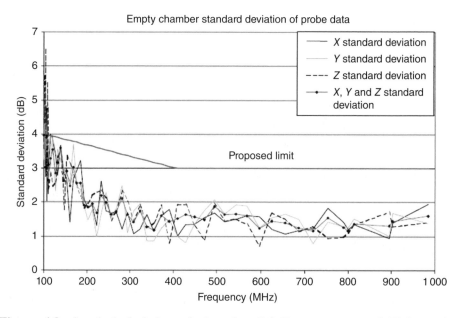

Figure 4.8   Standard deviation of data for *E*-field components of Eight probes. Source: IEC 61000-4-21 ed. 2.0 [3]. © 2011 IEC Geneva, Switzerland.

have used the IEC standard requirements, given in Table 4.5, where the specified field uniformity is achieved over a volume defined by an eight-location calibration data set. Another example can be found from Appendix D where the Liverpool Chamber was used and much finer frequency and angular resolution was used which gave more accurate and reliable results. The LUF of the Liverpool Chamber depends on the stirrer. For the latest stirrer, the LUF is about 140 MHz.

Generally speaking, the chamber mode density and the effectiveness of the mechanical stirrer determine the LUF which is generally accepted to be the frequency at which the chamber meets operational requirements. This frequency usually occurs at a frequency slightly above three times the first chamber resonance. In practice the chamber size, stirrer effectiveness, the chamber quality factor as well as the field uniformity and working volume determine the LUF – all these parameters are interlinked. For example, if the working volume is reduced, the LUF could be reduced and the field uniformity could be improved; if the field uniformity requirement is reduced, the LUF could be reduced and the working volume could be increased. Thus carefully calibration of the RC is essential for the use of RC for EMC testing.

4.4.2.3 Chamber *E*-field

For EMC measurements, especially for the immunity testing, it is important to know the EUT exposed field level. In an RC, this is the chamber *E*-field which should be the same in the EUT area.

The 'expected' amplitude of the chamber *E*-field during the calibration is simply the average of the 24 maximum probe readings (the mean of the maximums) at the corner locations of the EUT area. The 'expected value' is the value to which the chamber is calibrated in the EUT area. It is also possible to estimate the chamber *E*-field (E_{Est}) based on the reference antenna measurements [3]:

$$E_{Est} = \left\langle \frac{8\pi}{\lambda} \sqrt{5 \frac{P_{MaxR}}{\eta_{rx}}} \right\rangle_n \qquad (4.1)$$

where λ is the wavelength; P_{MaxR} is the maximum received power (in W) over the given number of stirrer steps at an antenna location or orientation; η_{rx} is the receiving antenna efficiency; n is the number of antenna locations and orientations and it should be at least 24.

Equation 4.1 was derived using methods similar to those used to derive expressions for the mean field in Ref. [16], gives an estimate of the chamber *E*-field based on the maximum readings from the reference antenna averaged over the number of antenna locations and orientations (n).

For all measurements, it is assumed that the forward input power is the same for all data collected. If so, then the data can be normalized after taking the average of the probe readings. If not, then the probe readings need to be normalized to the input power, which corresponds to that probe reading. Normalizing the *E*-field to the chamber input power is done by dividing the probe reading by the square root of the input power. This can also be done for the estimated *E*-field based on the reference antenna.

It is recommended that a crosscheck be performed by comparing the expected *E*-field measured by the probes and the expected *E*-field estimate based on the eight antenna measurements. According to the IEC standard, any discrepancies greater than ±3 dB between the probe- and antenna-based measurements should be resolved, but how to resolve it was not mentioned in the literature. The possible solutions are (i) to increase the number of stirrer steps (to reduce the uncertainty) and (ii) to reduce the EUT area (to improve the uniformity).

It was noted that significant disagreement at the lower frequencies is expected. This is primarily due to loading caused by the transmitting and receiving antennas. For this reason, the agreement between the two methods is not expected at frequencies where the difference between the chamber input power and the measured maximum received power from the reference antenna is 10 dB or less. Unlike in an AC, the fields inside an RC are of statistic meanings with certain uncertainty which is possibly the main reason why the RC is still not widely used in practice.

4.4.2.4   Loading Effects

When an EUT is placed into an RC, there is a possibility that the EUT will 'load' the chamber. If the EUT loads the chamber, the energy absorbed by the EUT is no longer available to generate the desired environment, although the general statistic features of an RC are maintained. For this reason, the chamber input power needs to be increased to compensate for this loading. This is another major difference from the testing in an AC where the reflected waves are absorbed by radio absorbing materials on the boundaries. The presence of the EUT does not change the incoming field to the EUT in the radiated immunity testing or the received radiated field from the EUT in the radiated emission measurement.

According to IEC 61000-4-21, prior to performing any test, a check for loading effects must be made. This is done by measuring the mean power received by the reference antenna, which is placed inside the EUT area as well (the selection of the location should be careful and representative), for the same number of stirrer steps used to perform the calibration with the EUT in place. The data from this single measurement are then compared to the eight measurements from the calibration. If the mean received power measured with the EUT in place is comparable with the mean field measured during the calibration (i.e. it is not greater or less than the calibration data), then the chamber is considered to have not been loaded by the EUT. If the measurement exceeds the uniformity of the mean field measured during calibration, a correction factor will be required when calculating the input power necessary to generate the desired test field. This factor is referred to as the chamber loading factor (CLF). The CLF is obtained by taking the ratio between the measurement value taken with the EUT in place and the mean or 'expected value' from the eight measurements taken during the calibration. To determine the limit to which a chamber may be loaded, an evaluation must be performed to assess the field uniformity under severe loading conditions as discussed in Annex B in IEC 61000-4-21 [3].

The loading effects will affect both the radiated emission and immunity tests.

4.4.3   Radiated Immunity Tests

Once the RC is properly characterised, one is ready to conduct radiated EMC measurements inside the chamber. In this section, we introduce and discuss the radiated immunity test, which is very different from the conventional radiated immunity test inside an anechoic or semi-anechoic chamber. A major advantage of using an RC for radiated immunity testing is that high level EM fields can be generated efficiently without using expensive high-power amplifiers since the RC has conductive boundary conditions and the power injected into the chamber is fully utilised for the test with little absorption. Again, we are going to use the IEC 61000-4-21 [3] as the major reference and most of the information here is from its Annex D. The test should follow the following six steps.

Step 1: Test set-up
The typical test set-up is shown in Figure 4.7. The equipment layout should be representative of the actual installation. The EUT should be at least $\lambda/4$ from the chamber walls at the LUF of the chamber. EUTs designed for tabletop operation should be located $\lambda/4$ from the chamber floor. Floor standing EUTs should be supported 10 cm above the floor, in the area beneath the uniform volume, by a low loss dielectric support. The layout of the test equipment and cables should be described in the test report.

The transmitting antenna should be in the same location as used for calibration. The transmitting antenna should not directly illuminate the EUT or the receiving antenna. Directing the antenna into the corners of the chamber is a recommended configuration (Note: this is also dependent on the chamber design, we find that the transmitting antenna directed to the stirrer is actually better than directed into a corner. It is important to ensure that there is no direct path or a strong reflected path from the transmitting antenna to the EUT). Appropriate modes of operation, software installation and stability of the EUT, test equipment and all monitoring circuits and loads should be established.

Step 2: Calibration
Prior to collecting data a check should be performed to determine if the EUT and/or its support equipment have adversely loaded the chamber. This check will be performed as outlined in Section 4.4.2.4. If mode-stirred procedures are used, care should be taken to ensure that the issues associated with stirring (such as the response time of the EUT and the rotation rate of the stirrer) are adequately addressed.

Step 3: Determining chamber input power requirements
Determine the chamber input power, P_{Input} (W), required meeting the test requirements for the electric field intensity using the equation:

$$P_{\text{Input}} = \left[\frac{E_{\text{Test}}}{\langle E \rangle_{24 \text{ or } 9} \times \sqrt{\text{CLF}(f)}} \right]^2 \qquad (4.2)$$

where P_{Input} is the forward power in Watts into the chamber to achieve the desired field strength for immunity tests; E_{Test} is the required field strength in V/m and CLF(f) is the chamber loading factor as a function of frequency f as mentioned in Section 4.4.2.4.

$\langle E \rangle_{24 \text{ or } 9}$ is the average of the normalized E-field from the empty chamber calibration. It will be necessary to interpolate (linear interpolation) between the calibration frequency points (calibration at a finer step interval is also an option).

CAUTION: RF fields can be hazardous. Observe applicable national RF exposure limits.

Step 4: Selecting frequency sweep/step rates/intervals

Frequency sweep or step rates should be selected with consideration of EUT response time, EUT susceptibility bandwidths and monitoring test equipment response time. The scan rate selected shall be justified by this criterion, and documented in the test report. Unless otherwise specified by the test plan, the following guidance will be used for selecting test frequencies.

For test equipment that generates discrete frequencies, the minimum number of test frequencies shall be 100 frequencies per decade. The test frequencies shall be logarithmically spaced. As an example (above 100 MHz), a formula which can be used to calculate these frequencies in ascending order is as follows:

$$f_{n+1} = f_n \times 10^{1/(N-1)} \tag{4.3}$$

where n is an integer, between 1 and N; N is the number of frequency samples and f_n is the nth test frequency (f_1 is the starting frequency and f_N is the end frequency).

For example, if the starting frequency is 150 MHz and the number of frequency samples is 100, we have:

$f_1 = 150.00\,MHz;$
$f_2 = 153.53\,\text{MHz};$
$f_3 = 157.14\,\text{MHz};$
$f_4 = 160.84\,\text{MHz};$
$\ldots$

The dwell time at each test frequency should be at least 0.5 s, exclusive of test equipment response time and the time required to rotate the tuner (to a full stop). Additional dwell time at each test frequency may be necessary to allow the EUT to be exercised in appropriate operating modes and to allow for the 'off time' during low frequency modulation. At least two full cycles of modulation will be applied. For example, if the applied modulation is a 1 Hz square wave modulation, the dwell time should not be less than 2 s. The dwell time selected is justified based on EUT and test equipment response time, as well as applied modulation, and documented in the test report.

For test equipment that generate a continuous frequency sweep, the minimum (fastest) sweep rate shall be equal to the number of discrete frequencies per decade multiplied by the dwell time, that is, 100 discrete frequencies per decade times 1 s dwell time equals 100 s per decade sweep rate. The fastest sweep rate shall be used only when the EUT and associated test equipment are capable of fully responding to the test stimulus. If it cannot be verified that the EUT can adequately respond to the swept stimulus, then discrete frequencies shall be used unless a sweep rate is specified by the product committee. The use of stirring in conjunction with swept frequency testing is discouraged.

NOTE: Additional test frequencies should be included for known equipment response frequencies, such as image frequencies, clock frequencies, etc. Specific test, manufacturer or government/regulatory requirements may list a specific scan rate or frequency interval(s) that takes precedence.

Step 5: Performing the test

The test can be performed using either mode-tuned or mode-stirred procedures. For mode-tuned operation, use the minimum numbers of steps as indicated by the chamber calibration. The tuner should be rotated in evenly spaced steps so that one complete revolution is obtained per frequency. If mode-stirred procedures are used, it is ensured that the EUT is exposed to at least the number of samples as the calibration equipment was during calibration. Ensure that, for either procedure, the EUT is exposed to the field level for the appropriate dwell time. This is particularly important for mode-stirred operation.

It should be noted that the chamber calibration allows the number of steps to be reduced to 12 if the data indicates that an acceptable chamber performance can be achieved. Monitor and record maximum received power (P_{MaxR}) and average received power (P_{AveR}) with the receiving antenna used in the calibration of each frequency band to ensure that the required field strength is being generated. Use P_{AveR} to ensure that the chamber loading has not changed from the calibration. Differences greater than 3 dB in P_{AveR} from that obtained in Step 2 should be resolved. P_{MaxR} may also be used to estimate the peak E-field generated.

Monitor and record the average values of P_{Input} and $P_{Reflected}$. Variations in P_{Input} over a tuner rotation greater than 3 dB should be noted in the test report.

Modulate the carrier as specified in the test plan. When modulation is applied, ensure that the peak amplitude complies with the definitions of the test plan.

Scan the frequency range to the upper frequency limit using the appropriate antennas and modulations.

It should be noted that when using pulse modulation, ensure that the time response of the chamber is fast enough to accommodate pulsed waveform testing. Linear interpolation between calibration points will be required.

Step 6: Test report

The test report, which is normally required for any EMC test, should include the cable layout and positioning of the EUT relative to the cables and a diagram of the test set-up and/or photos. In addition, the report should include the following parameters for each test frequency, in addition to the reporting requirements related to the EUT:

1. maximum received power from the receiving antenna used to monitor the field in the chamber
2. mean received power from the receiving antenna used to monitor the field in the chamber
3. forward power delivered to the chamber transmitting antenna
4. reflected power from the chamber transmitting antenna
5. variations in forward power during the data collection period greater than 3 dB
6. differences greater than 3 dB between field levels based on chamber input power and that calculated using the methods mentioned in Section 4.4.2.3 which could not be resolved

The measurement method and procedures in other standards using RCs are similar to the six steps presented here.

4.4.4   Radiated Emission Measurements

This section covers the use of RCs to make radiated power measurements of emissions (both intentional and/or unintentional) but it does not cover all of the nuances of emissions measurements. Information on measurement apparatus can be found in CISPR 16-1-1 [12]. In general the information found in CISPR 16-1-1 applies without modification when making measurements using an RC.

There are two exceptions which require additional consideration:

1. the distortion of short duration pulses (typically defined as less than 10 μs) by the chamber quality factor or 'Q' and
2. the apparent amplitude variation of the emitted signal due to motion of the mechanical stirrer.

Determination of suitable chamber Q (i.e. time constant) can be found in Chapter 3. Effects of the stirrer should be considered when selecting the dwell time or rotation rate and when selecting the type of detector to be used.

Step 1: Test set-up
Test set-up information contained in CISPR 16-2-3 [13] applies to RC testing. A typical test set-up is shown in Figure 4.7. The only additional requirements are that the EUT shall be at least $\lambda/4$ from the chamber walls and floor standing EUTs shall be supported 10 cm above the floor by a low loss/low permittivity dielectric support. The use of a ground plane is allowed if necessary for proper operation of the EUT. In addition, the

need to manipulate the position of interface cables is eliminated and the support table should be non-absorbing as well as non-conductive.

The transmitting (Tx) antenna (used during chamber calibration to check the chamber for loading) should remain in the chamber at the same location as used for calibration. The Tx antenna will not directly illuminate the EUT or the receiving (Rx) antenna. The Rx antenna will not be directly illuminated by the EUT (i.e. the Rx antenna should not be directed at the EUT). Directing the antennas into the corners of the chamber is an optimum configuration. (Note: as we mentioned earlier, it is dependent on the chamber design. In some case it would be better to direct the antenna at the stirrer.) Establish software installation, modes of operation and stability of the EUT, test equipment and all monitoring circuits and loads.

Step 2: Calibration
Prior to collecting data a check is performed to determine if the EUT and/or its support equipment have adversely loaded the chamber. This check is performed as outlined in Section 4.4.2.4. If mode-stirred procedures are used, care is taken to ensure that the issues associated with stirring are adequately addressed. Once the loading check has been performed, the transmitting antenna shall be terminated with a characteristic impedance equivalent to the RF source used during calibration.

Step 3: Radiated emission test procedures
The test is performed using either mode-tuned or mode-stirred procedures. Ensure that for either procedure the EUT is sampled by at least the number of samples as the calibration equipment was during calibration. For mode-tuned operation, use the minimum number of samples as indicated by the chamber calibration. The stirrer should be rotated in evenly spaced steps so that one complete revolution is obtained per frequency. If mode-stirred procedures are used, it is ensured that the EUT emissions are sampled with at least the number of samples collected during the chamber calibration. As with mode-tuned, the mode-stirred samples should be uniformly spaced over one complete stirrer rotation.

Ensure for either procedure that the EUT is monitored at each sample for a time period sufficient to detect all emissions (see CISPR 16-2-3 for guidance on receiver scan times). This is particularly important for mode-stirred operation. Mode-stirred procedures should only be applied for unmodulated signals using a peak detector. Due to the amplitude variation of the received signal caused by the motion of the stirrer, testing time will usually be increased if a peak detector is to be used. Mode stirring is not applicable when using an average or other weighting detector.

For modulated emissions, radiated mean power within the measurement bandwidth will be measured, if an root mean square (RMS) detector is used. If the emissions spectrum is wider than the measurement bandwidth, the total radiated mean power can be measured by integrating the mean power spectral density over the emission spectrum.

Monitor and record P_{MaxR} and/or P_{AveR} as specified in the test plan with the Rx antenna used in the calibration of each frequency band.

CAUTION: To get an accurate measure of P_{AveR}, the noise floor of the receiving equipment should be at least 20 dB below P_{MaxR}.

Scan the frequency range to the upper frequency limit using the appropriate antennas and bandwidths. The scan time for this procedure should be as specified in the test plan.

Step 4: Determining radiated power

Measuring the amount of power received by the receiving antenna and correcting for chamber losses can determine the amount of RF power (within the measurement bandwidth) radiated by a device placed in the chamber. The power radiated from a device can be determined using either average or maximum received power. Equation 4.4 is used for average-received-power-based measurements and Equation 4.5 is used for maximum-received-power-based measurements [3].

$$P_{Rad} = \frac{P_{AveR} \times \eta_{Tx}}{CCF} \tag{4.4}$$

$$P_{Rad} = \frac{P_{MaxR} \times \eta_{Tx}}{CLF \times IL} \tag{4.5}$$

where P_{Rad} is the radiated power from the device (within the measurement bandwidth); CCF is the chamber calibration factor and is calculated using (n is the number of antenna locations).

$$CCF = \left\langle \frac{P_{AveR}}{P_{Input}} \right\rangle_n \tag{4.6}$$

CLF is the chamber loading factor; IL is the chamber insertion loss and can be calculated using:

$$IL = \left\langle \frac{P_{MaxR}}{P_{Input}} \right\rangle_{8 \text{ or } 3} \tag{4.7}$$

8 probe/antenna locations are used if the frequency is below $10 f_s$ or 3 probe/antenna locations are used for this calculation.

P_{AveR} is the received power (within the measurement bandwidth) as measured by the reference antenna averaged over the number of stirrer steps; P_{MaxR} is the maximum power received (within the measurement bandwidth) over the number of stirrer steps and η_{Tx} is the antenna efficiency factor for the Tx antenna used in calibrating the chamber and can be assumed (if not known) to be 0.75 for a log-periodic antenna and 0.9 for a horn antenna (Note: these values may be smaller than some commercial antennas).

The advantage of using measurements based on average power is a lower uncertainty. The disadvantage is that the measurement system will have sensitivity of 20 dB lower than the measured P_{MaxR} to get an accurate average measurement.

Step 5: Estimating the free space (far) field generated by an EUT

The electric field strength generated by the EUT at a distance of R meter(s) can be estimated by using the equation:

$$E_{Rad} = \sqrt{\frac{D \times P_{Rad} \times \eta_0}{4\pi R^2}} \qquad (4.8)$$

where E_{Rad} is the estimated electric field strength generated by the EUT in V/m; P_{Rad} is the radiated power from (4.5) in W; η_0 is the intrinsic impedance of free space and is about 377 Ω; R is the distance from the EUT in metres and should be a sufficient distance to ensure far field conditions exist and D is the equivalent directivity of the EUT. A directivity of $D=1.7$ is often used as it represents the assumption that the EUT radiation pattern is the equivalent of a dipole radiator. It is recommended that a factor of 1.7 be used unless the product committee can supply a more appropriate value. Some research on device directivity can be found in such as [14].

The calculated disturbance field strength is not always compatible with measurement results given at the Open Area Test Sites (OATSs) or similar test sites. This compatibility, if required, is shown by specific procedures for EUT types or product groups.

It should be pointed out that, in reality, the free space model is not accurate since the radiated wave may be reflected by, such as the ground. The radiated field strength given by (4.8) should be modified by taking the reflected waves into account. But this predicted field is useful for the comparison with the field measured in an AC where there is no reflection from the environment.

Step 6: Test report

The test report should be produced at the end of the measurement and it should include the following parameters for each test frequency, in addition to the reporting requirements related to the EUT:

1. maximum received power from the receiving antenna if recorded
2. mean received power from the receiving antenna if recorded
3. power emitted by the EUT as defined in (4.4) or (4.5)
4. if estimated E-field is required to be reported then the assumed directivity used to calculate the E-field (see Eq. 4.6) is also to be reported
5. loading data as required by Step 3
6. cable layout and positioning of the EUT relative to the cables
7. diagram of the test set-up (e.g. photos)

4.4.5   An Example of Radiated Emission Measurements

To demonstrate the use of an RC for EMC measurements, an all-in-one computer and a WiFi router (Huawei H533), as shown in Figure 4.9, have been chosen as the EUT for radiated emission measurements. In this case the EUT were intentional radiators around 2.45 GHz, but not in other frequency bands. The computer and router formed a communication system which could operate both inside and outside the RC. The Liverpool University RC of dimensions $5.8 \times 3.6 \times 4\,m^3$ was employed. The LUF was about 150 MHz for the selected EUT area of $2 \times 1.5 \times 2\,m^3$.

Before the formal measurement, we checked the emission from the EUT over the frequency range from 150 MHz to 3 GHz, there was very little emission at other frequencies except around 2.45 GHz. This indicated that the EUT were of good quality and should have passed EMC tests as expected. Thus this demonstration was to measure radiated emissions around 2.45 GHz and the results will be used in the next section for comparing the results. In order to ensure the signals were radiated at fixed values, both the computer and router were turned on and in communication mode during the measurement. The measurement frequency range was from 2.3 to 2.6 GHz. Two commercial broadband antennas were used for this measurement: Antenna 1 was a log-period antenna 9143 from 0.3 to 5 GHz with the antenna radiation efficiency of about 85%, whereas Antenna 2 was a double-ridged horn antenna from 2 to 18 GHz with the antenna radiation efficiency of about 95% over the band of interest. A high performance Spectrum Analyser (SA) was selected as the receiver. A digital signal generator was used as the reference signal generator.

Step 1: Test set-up
The measurement set-up is shown in Figure 4.10. The transmitting/reference antenna, Antenna 1, remained in the chamber at the same location as used for calibration.

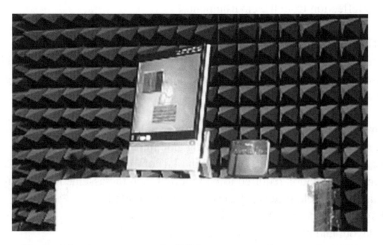

Figure 4.9   A computer and a WiFi router were chosen as the EUT.

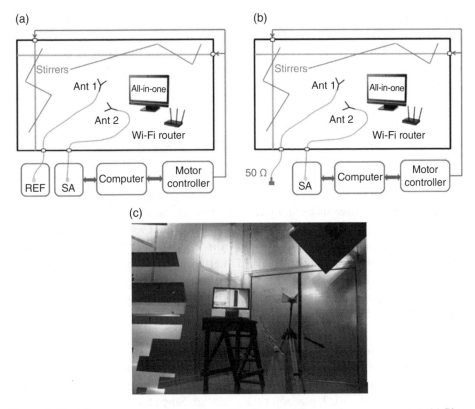

Figure 4.10   Measurement set-up. (a) For calibration. (b) For emission measurement. (c) Photo of the EUT and antennas inside the RC.

This antenna was directed to one of the stirrers and not illuminating the EUT or the receiving (Rx) antenna. The Rx antenna, Antenna 2, was in the EUT area and directed to a corner, not directly illuminated by the EUT – this was why a directional horn antenna was used in this case.

Step 2: Calibration

Prior to collecting data a check was performed to determine if the EUT and/or its support equipment have adversely loaded the chamber. Mode-tuned operation is selected at a step size of 1 degree, which means 359 samples for one complete revolution per frequency. As shown in Figure 4.10a, the signal generator was connected to Antenna 1 and signals were received by the SA using Antenna 2. At 2.45 GHz, the supplied power from the reference source was $P_{ref}=0$ dBm (1 mW), the received average power at the SA without the EUT was $P_{refNoEUT}=-38$ dBm – the loss was mainly caused by the chamber and two cables (2.5 and 3.5 dB, respectively).

The received power at the SA with the EUT was $P_{\text{refEUT}} = -40$ dBm, which means the additional loss introduced by the EUT was 2 dB which was acceptable.

After the chamber loading was checked and accepted, the transmitting antenna was terminated with a load of characteristic impedance (50 Ω) equivalent to the RF source used during calibration as shown in Figure 4.10b.

Step 3: Radiated emission test procedures

The test was performed using mode-tuned procedures. The EUT was sampled by 359 steps for one complete revolution per frequency. Monitor and record received power P_{R} as specified in the test plan with the Rx antenna used in the calibration of each frequency. A typical received power at one step is given in Figure 4.11a and the averaged power over one complete revolution is shown in Figure 4.11b. These received signals were from both the computer and the router which used IEEE802.11b standard. The channel allocations are shown in Figure 4.12: there are 14 channels and each has 22 MHz bandwidth. To avoid interference, the computer and the router had used different channels without overlap: one used Channel 6 and the other used Channel 11. The radiated powers were slightly different as we can see clearly from Figure 4.11.

Step 4: Determining radiated power

Measuring the amount of power received by the receiving antenna and correcting for chamber losses can determine the amount of RF power (within the measurement bandwidth) radiated by the EUT placed in the chamber. Since the radiated power is well above the noise floor, we use the average received power which has a smaller uncertainty, thus Equation 4.4 is employed for this case

$$P_{\text{Rad}} = \frac{P_{\text{AveR}} \times \eta_{\text{Tx}}}{\text{CCF}} = P_{\text{AveR}} \times \eta_{\text{Tx}} \times \frac{P_{\text{ref}} \times L_{\text{cable1}} \times \left(1 - |S_{11}|^2\right)}{P_{\text{refEUT}}} \tag{4.9}$$

where L_{cable1} is the loss factor of Cable 1 which is connected to Antenna 1; S11 is the reflection coefficient of Antenna 1; P_{AveR} and P_{refEUT} are received average power from the EUT in Step 3 and received average power from the reference source with the presence of the EUT in Step 2, respectively. They all are functions of frequency. $P_{\text{ref}} = 0$ dBm (1 mW) is the source power, and $P_{\text{ref}} \times L_{\text{cable1}} \times (1 - |S_{11}|^2)$ is the forward power to the Tx antenna averaged over one stirrer rotation.

The final radiated power from the EUT using (4.9) is obtained and shown in Figure 4.13, which has excluded the cable and chamber losses and is the absolute value from the EUT. The radiated spectrum power density at Channel 6 is about -2 dBm/MHz, while the spectrum power density at Channel 11 is about 0 dBm/MHz which are within the specs.

(a)

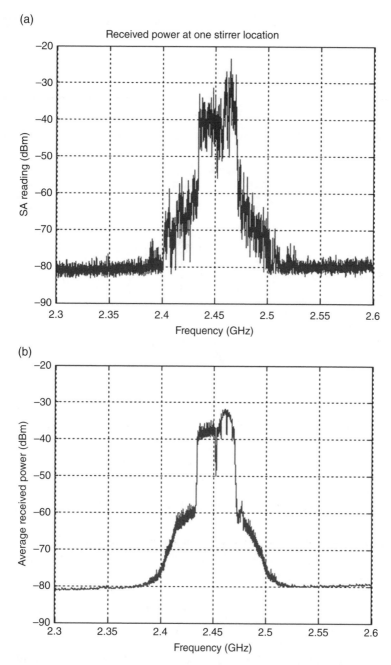

Figure 4.11   Received power. (a) At one step and (b) the average.

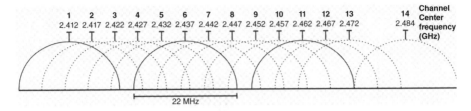

Figure 4.12   IEEE802.11b channels http://en.wikipedia.org/wiki/List_of_WLAN_channels

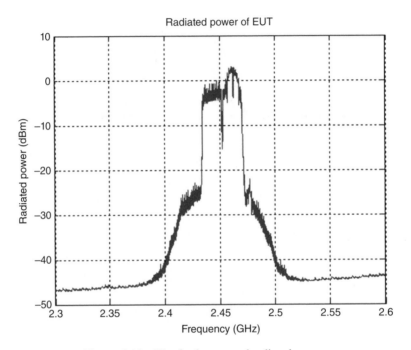

Figure 4.13   The final measured radiated power.

Step 5: Estimating the free space (far) field generated by the EUT

The field strength generated by the EUT at a distance of R meter(s) can be estimated by using Equation 4.8, that is:

$$E_{Rad} = \sqrt{\frac{D \times P_{Rad} \times 30}{R^2}}$$

where P_{rad} is obtained from (4.9) and the directivity D of the EUT in our case is unknown, which is a drawback of using the RC, and it can only be estimated as say $D = 1.7$, as recommended. This should be a reasonable estimate since the router has a dipole-type antenna. The directivity of a half-wavelength dipole is 1.64.

The measurement time for this was about 5 h (due to the fine step of 1° and 10 001 frequency points used for more accurate results) using a computer-controlled automatic measurement system. The data processing was relatively straightforward.

4.5 Comparison of Reverberation Chamber and Other Measurement Facilities for EMC Measurements

Since the introduction of using an RC for EMC measurements, there has been a question about if the results obtained from the RC are comparable with that from other established measurement facilities (such as the OATS, anechoic and semi-anechoic chamber). Many people have tried to deal with this question using analytical, numerical and experimental approaches [15–17]. One distinct difference, which bothers some people, is that the measured result in an RC is only meaningful in term of the statistic feature (typically the average value over a revolution of the stirrer), while the measured result in other measurement facilities is deterministic and there is no need for averaging. This is something we have to accept and get used to it since an RC is such a facility that a single measurement result is not representative enough and we need to use the averaged result. Smooth results can be obtained if frequency averaging is applied.

For radiated emission measurements, there have been many publications, since this is an area which is relatively easy to obtain results for comparison. In Ref. [15], Harrington has presented an interesting study on radiated emissions from a dipole antenna and a simple box EUT measured in free space (OATS), inside gigahertz transverse electromagnetic (GTEM) cells and RCs. A typical comparison of the results is given in Figure 4.14 where a 3 m OATS was used and results were compared with that

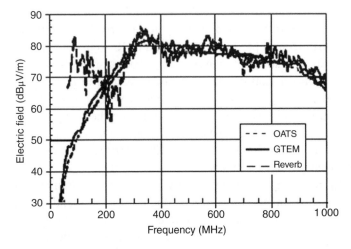

Figure 4.14 A typical measured and predicted OATS 3 m horizontal polarisation electric field response. Source: Harrington [15]. Reproduction courtesy of the IEEE.

from a GTEM cell and an RC (calculated from the measured total radiated power). Very good agreement was shown between measured results. Although the RC and GTEM emission field strengths exhibited more ripple than in an OATS, the average predicted field levels were comparable between the various facilities. For some EUTs, six or more positions might be required in a GTEM cell to ensure a full capture of the total radiated power, whereas a properly operating reverberation would collect all power without any EUT rotations (instead, the stirrer rotations were required which is an advantage of an RC). Based on these results, it was concluded that both the RC and GTEM cell were capable of providing useful pre- and even full-compliance emission testing, at least within some error proportional to the maximum and minimum response envelope.

For the case presented in Section 4.4.5, the emissions from the EUT were also measured inside an AC as shown in Figure 4.15a. The measured radiated electric field is compared with the one estimated using the radiated power measured from the RC as discussed in Section 4.4.5. A comparison of the results is given in Figure 4.15b. We can see that they are in reasonable agreement – since the radiation from the EUT, especially the PC, did not have an omnidirectional radiation pattern and the measured E-field in this figure was just for one direction in the AC, while the calculated E-field from the RC result has no directional information, we therefore should not expect that they are perfectly matched. The difference between the noise floor levels in the two environments was due to the equipment setting and dynamic range. If we used different settings, the noise floor could be tuned to about the same level for both cases.

For radiated immunity and susceptibility tests, a considerable amount of work has also been done. A comparison of radiated susceptibility test results using the RC and the semi-anechoic chamber were presented in Ref. [16] where the calibration and test procedures were the same as specified in Section 20 of DO-160D, Change One [8]. An artifact that was representative of many typical poorly designed EUTs was used for the test. Both test methods produced high levels of RF signal being coupled onto the pickup loop inside the artifact as measured by a power meter outside the chambers. In analysing the data, it appears that both methods tend to produce the same peaks across the frequency range tested (between about 400 MHz and 18 GHz), but that the RC tends to 'pull up' the valleys, mostly eliminating the many points of minimum coupling found in the semi-anechoic chamber test results. The author considered that the RC method did a better job of coupling RF energy into a 'typical' avionics box, across the frequency range of interest.

In Ref. [17], the comparison between RC and AC for an immunity test was considered. A two-wire transmission line was adopted as a DUT, and the current induced on the line by the external field was monitored to construct a possible susceptibility profile. Both the averaged current over stirrer rotation and the maximum induced current were considered to show how robust the RC measurement was against the positioning and the orientation of the DUT inside the chamber. As expected from

(a)

(b)

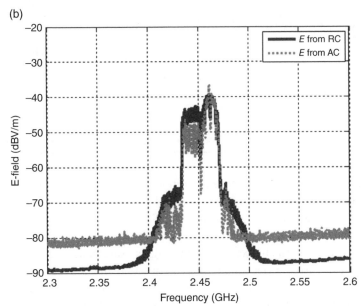

Figure 4.15 Radiated emission measurements. (a) EUT in an AC. (b) Comparison of measured electric fields in an RC and an AC.

theory, the maximum value was related to the average one by means of the number of the independent positions of the stirrer.

A comparison between the RC and AC measured data is presented in Figure 4.16. The induced current was normalised to the field in both cases. The RC results were reported

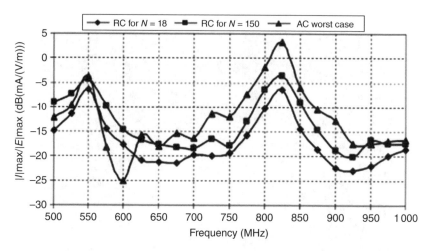

Figure 4.16 Comparison between RC and AC maximum induced current on the DUT. Source: De Leo and Primiani [17]. Reproduction courtesy of the IEEE.

Table 4.6 Comparison between RC and AC immunity tests [17].

Parameters	Reverberation chamber	Anechoic chamber
Uncertainty (expanded, $k = 2$)	±3.3 dB in the case	±2.6 dB in the case
Power required for 100 V/m	11 W (average field) 1.6 W (Max field)	300 W ($G_T = 10$ and $d = 3$ m)
Relative cost (only chamber)	1	7–10
Testing time	$18(12) \times N_{freq} \times$ Dwell time	$2 \times 4(8) \times N_{freq} \times$ Dwell time

for two values of the stirrer independent positions ($N = 18$ and $N = 150$). The result with more independent positions was closer to the AC in worst case. It was noted that it was possible to lose some worst case situations in the RC, related to a particular incident field polarisation and incoming direction, and the underestimation of the induced current depended upon the chosen number of stirrer positions. An adequate safety margin should be considered in carrying out immunity tests in RC if the worst case AC is sought, depending upon device directivity and stirrer-independent positions. On the other hand, it must be said that for complex objects, it is difficult to capture the worst case in AC (corresponding to a particular polarisation and field incoming direction) unless a great number of measurements are done, with a very time-consuming procedure, to investigate all possible angles of incidence. Moreover, in particular situations, automotive radiated immunity testing, for example, it is impossible to investigate some incoming directions, whereas in the RC, all directions, although in a statistical sense, are used. Some other important issues were also discussed and summarised in the paper [17]. A comparison table is reproduced in Table 4.6. The uncertainty in the RC is generally higher than that in the AC, but the power required and cost for an RC are normally much smaller than it is in an AC. The testing time is dependent on a number of parameters. For a standard test inside an AC, the antenna is expected to change its height, typically from 1 to several metres, but there is no such requirement for the test inside an RC. It is now widely accepted that radiated EMC testing inside an RC is faster than it is inside an AC [18], but unlike inside an AC, there is no information on the direction and polarisation of the wave/field inside an RC.

4.6 Conclusions

This chapter has shown how to use an RC for EMC measurements. Detailed measurement set-up, procedures, relevant theory and equations have also been presented. The most important standard IEC 61000 Part 4-21 has been used as the major reference and some testing examples have been given to illustrate how to conduct radiated EMC measurements. The comparison of the RC and other conventional EMC measurement facilities has been conducted and their advantages and disadvantages have been identified.

Acknowledgements

The authors thank the International Electrotechnical Commission (IEC) for permission to reproduce information from its International Standard IEC 61000-4-21 ed. 2.0 (2011). All such extracts are copyright of the IEC, Geneva, Switzerland. All rights reserved. Further information on the IEC is available from www.iec.ch. IEC has no responsibility for the placement and context in which the extracts and contents are reproduced by the authors, nor is the IEC in any way responsible for the other content of accuracy therein.

References

[1] H. A. Mendes, 'A new approach to electromagnetic field strength measurements in shielded enclosures', Wescon Technical Papers, Los Angeles, August 1968.
[2] P. Corona, G. Latmiral, E. Paolini and L. Piccioli, 'Use of reverberating enclosure for measurements of radiated power in the microwave range', *IEEE Transaction on Electromagnetic Compatibility*, vol. 18, pp.54–59, 1976.
[3] IEC 61000 Part 4-21: testing and measurement techniques – Reverberation chamber test methods, ed. 2, 2011.
[4] 'American National Standard Dictionary for Technologies of Electromagnetic Compatibility (EMC), Electromagnetic Pulse (EMP) and Electrostatic Discharge (ESD) (Dictionary of EMC/EMP/ESD Terms and Definitions)', ANSI C63.14-1998, 1998.

[5] DIRECTIVE 89/336/EEC OF THE EUROPEAN PARLIAMENT AND OF THE COUNCIL, 1989, which is now replaced by 2004/108/EC. http://eur-lex.europa.eu/LexUriServ/LexUriServ. do?uri=OJ:L:2004: 390:0024:0037:EN:PDF (accessed 4 August 2015).

[6] M.L. Crawford, 'Generation of standard EM fields using TEM transmission cells', *IEEE Transaction on Electromagnetic Compatibility*, vol. 16, pp. 189–195, 1974.

[7] A. Nothofer, D. Bozec, A. Marvin and L. McCormack, 'Measurement Good Practice Guide, No. 65: The Use of GTEM Cells for EMC Measurements', National Physical Lab (NPL), UK, 2003. http:// site.yorkemc.co.uk/assets/YorkEMC-NPL_gtem-good-practice.pdf (accessed 4 August 2015).

[8] RTCA DO 160D. http://everyspec.com/MISC/download.php?spec =RTCA_DO-160D.048944.pdf (accessed 4 August 2015).

[9] L. R. Arnaut, 'Time-domain measurement and analysis of mechanical step transitions on mode-tuned reverberation chamber: Characterisation of instantaneous field', *IEEE Transaction on Electromagnetic Compatibility*, vol. 49, pp. 772–784, 2007.

[10] V. Rajamani, C. F. Bunting and J. C. West, 'Stirred-mode operation of reverberation chamber for EMC testing', *IEEE Transaction on Electromagnetic Compatibility*, vol. 61, pp. 2759–2764, 2012.

[11] 'EN 61000-4-21:2003 Electromagnetic compatibility (EMC). Testing and measurement techniques. Reverberation chamber test methods,' ed, 2003.

[12] CISPR 16-1-1 (Measuring apparatus), ed. 3.2, IEC, June 2014.

[13] CISPR 16-2-3 (Radiated disturbance measurements), Am2 ed. 3.0, IEC, March 2014.

[14] P. Wilson, G. Koepke, J. Ladbury and C.L. Holloway, 'Emission and immunity standards: replacing field-at-a-distance measuremnt with a total-radiated power measurements', *2001 IEEE International Symposium on Electromagnetic Compatibility, 2001*. EMC, August 2001, Motreal.

[15] T. Harrington, 'Total-radiated-power-based OATS-equivalent emissions testing in reverberation chambers and GTEM cells', *IEEE International Symposium on Electromagnetic Compatibility, 2000*. EMC, August 2000, Washington, DC.

[16] E.J. Borgstrom, 'A comparison of methods and results using the semi-anechoic and reverberation chamber radiated RF susceptibility test procedures in RTCA/DO-160D', *2004 International Symposium on Electromagnetic Compatibility, 2004*. EMC 2004, August 2004, Washington, DC.

[17] R. De Leo and V. Primiani, 'Radiated immunity tests: reverberation chamber versus anechoic chamber results', *IEEE Transactions on Instrumentation and Measurement*, vol. 55, 1169, 2006.

[18] M. Hoijer, 'Fast and accurate radiated susceptibility testing by using the reverberation chamber', *2011 IEEE International Symposium on Electromagnetic Compatibility (EMC)*, August 2011, Long Beach, CA.

5

Single Port Antenna Measurements

In this chapter, we will discuss the characterisation of single port antennas in the Reverberation Chamber (RC). The main focus and aim of this chapter will be to determine the efficiency of single port antennas. This represents an antenna parameter to which the RC lends itself well to accurately characterising. The concept of antenna efficiency will be introduced before detailing all appropriate measurement equations, proofs and procedures, which will allow a user to undertake such measurements with confidence and accuracy. This aspect will be further reinforced by a discussion on potential pitfalls to avoid when undertaking such measurements.

The antenna performance results in this chapter will concentrate on a relatively new and popular research theme in the antennas field: the textile antenna. Although this will be the subject focus, the procedures and equations that are depicted throughout this chapter are generally transferrable to other types of single port antennas; conventional or otherwise.

A secondary aim and purpose of this chapter will be to show how the RC can be used to provide an accurate assessment of the 'installed' performance of antennas – that is taking into account more complete factors from a system perspective. It is to this aim as to why the textile antenna has been selected as the focus of this chapter as will be explained next. There are many important parameters to evaluate an antenna's performance, the radiation efficiency is one of the particular interests for RC measurements since it is very hard to obtain using conventional measurement methods, but is relatively easy using the RC. Thus the measurement of radiation efficiency is the focus of this chapter.

Reverberation Chambers: Theory and Applications to EMC and Antenna Measurements, First Edition.
Stephen J. Boyes and Yi Huang.
© 2016 John Wiley & Sons, Ltd. Published 2016 by John Wiley & Sons, Ltd.

5.1 Introduction

A growing interest is evident concerning body-centric wireless communication technology, aimed at providing solutions for a wide range of applications from the medical/healthcare industries, the consumer electronic industries, wearable technology for the fashion industry and the military sector to name but a few. A crucial component in the body-centric wireless communication 'chain' concerns the antenna device itself; the antenna should be ergonomically suitable for integration onto a human body. Furthermore, the antenna should not be obtrusive, it should be able to maintain flexibility, exhibit minimal degradation in terms of bandwidth, reflection coefficient and efficiency performance and be manufactured in a low-cost manner. To satisfy this broad criterion, the textile antenna has been developed and continues to receive ever-growing attention [1–9].

When communicating wirelessly in the proximity of a human body, the propagation channel is dependent upon the body condition, the human activity being performed, the antenna position, the immediate surrounding environment and any interaction between the human body and the antenna [10–12]. Therefore, the radio propagation channel in this (on-body) scenario directly includes the body effect and is not usually stationary.

As previously stated, the main parameter of interest concerns the radiation and total radiation efficiency performance of some newly designed textile antennas. This work results from a collaboration between two parties; the textile antennas are the product of the work performed by P. J. Soh and G. A. E. Vandenbosch [3, 4].

The RC measurement facility is chosen partly because of the statistically emulated environment it can offer, but also it will be shown that it can offer a low uncertainty measurement solution to the problem of measuring textile antenna efficiencies when acquired in conjunction with a human being. To explain further, it is inevitable that some human movements will be present during any measurement procedures due to breathing and so on. This could prove problematic in other facilities, but in the RC environment they could be tolerated as any movements should add to the randomness of the field inside the chamber which can be viewed positively. A further benefit exists with the RC in the fact that it is potentially relatively quicker and easier to start the measurement process.

A main experiment concerning antenna radiation efficiency analysis measured on live human beings in an RC was reported in Ref. [13]. The conclusions drawn from this piece of work indicated that RC results obtained with human test subjects was sufficiently representative of results acquired in anechoic chambers and also measured against tissue phantoms – this potentially validates the live human approach against phantoms often used in the field, and benchmarks the RC facility against alternative measurement facilities.

The results from Ref. [13] also indicated that the type of antenna plays a more dependent role in determining the overall efficiency performance, as opposed to the characteristics of different human subjects.

The work in this chapter will present findings to address information concerning the performance of textile antennas specifically designed for on-body communications; particularly when worn on-body. An assessment of movement-related bending effects on antenna performance will be provided as well as the effect of different body locations and proximity distances from the human on how the textile antenna performs.

This chapter will present detailed measurement procedures in the RC for both free space and on-body conditions, as well as the statistical effects of different human subjects on RC measurements. It will also chart information concerning the repeatability of such measurements. Furthermore, definite uncertainty deductions will also be discussed so that absolute confidence can be placed in antenna performance results when acquired on human beings.

5.2   Definitions and Proof: Antenna Efficiency

5.2.1   Radiation Efficiency

Generally, there are two components that make up the quantity of antenna efficiency: the *radiation efficiency* and the *total radiation efficiency*. For the quantity of radiation efficiency, the IEEE standard definition of terms can be referred to which describe the quantity as [14]:

The ratio of total radiated power to net power accepted by the antenna at its terminals.

The quantity of radiation efficiency therefore describes the ohmic losses that exist within a given antenna structure – losses that are due to factors such as the material characteristics of the antenna. In this particular quantity, the term *net* in the definition is crucial. This defines that any associated loss factors such as losses due impedance mismatch are *not taken into account* by this quantity. For example, if an antenna is poorly impedance matched, say only 1% of power is able to be delivered to the antenna, if the antenna radiates all of that 1% of power, then by definition it will be 100% efficient in terms of its radiation efficiency.

We will now describe how this quantity can be calculated using the RC test methods. Standard techniques using the RC to measure the efficiency of antennas rely on the use of a reference antenna – that is, an antenna which has known performance characteristics.

Recently derived techniques have been published which seek to relax this constraint; these can be found in Chapter 7. However, in this chapter we will seek to derive and perform the technique around the standard means of measuring. This means that we will measure two separate antennas on the receiving side; one is the antenna with unknown efficiency which is denoted as Antenna Under Test (AUT), the other is a reference antenna with known value which will be denoted as REF.

Beginning the process of using the RC for antenna efficiency measurements, and assuming the use of a reference antenna and a Vector Network Analyser (VNA) transmitting from Port 1 we can state that:

The power transmitted by Port 1 of the VNA (say, P_{Source}) will first encounter the reflection by the fixed transmitting antenna (due to impedance mismatch) with voltage reflection coefficient S_{11} and radiation efficiency η_F, Thus the power radiated into the chamber (neglecting the cable and connector loss for the whole system) is:

$$P_{\text{Chamber}} = P_{\text{Source}}\left(1 - |S_{11}|^2\right) \cdot \eta_F \tag{5.1}$$

Part of this power is reached to the receiving antenna (which is either the reference antenna or the AUT) and it can be expressed as:

$$P_{\text{in}} = P_{\text{Chamber}} - P_{\text{Loss}} \tag{5.2}$$

where P_{Loss} is the power loss due to the chamber ohmic loss and power spreading.

The power at the receiving antenna is assumed to be the same for both REF and AUT antennas and it is partially lost (due to antenna efficiency) and partially reflected (due to mismatch between the antenna and cable connector), thus the power eventually transmitted to VNA Port 2 is:

$$P_{\text{R\_i}} = P_{\text{in}} \cdot \eta_i \cdot \left(1 - |S_{22i}|^2\right) \tag{5.3}$$

Thus we have:

$$|S_{21\text{REF}}|^2 = \frac{P_{\text{R\_REF}}}{P_{\text{Source}}} = \frac{P_{\text{in}} \cdot \eta_{\text{REF}} \cdot \left(1 - |S_{22\text{REF}}|^2\right)}{P_{\text{Source}}} \tag{5.4}$$

And:

$$|S_{21\text{AUT}}|^2 = \frac{P_{\text{R\_AUT}}}{P_{\text{Source}}} = \frac{P_{\text{in}} \cdot \eta_{\text{AUT}} \cdot \left(1 - |S_{22\text{AUT}}|^2\right)}{P_{\text{Source}}} \tag{5.5}$$

Dividing (5.4)/(5.5), we obtain:

$$\eta_{\text{AUT}} = \left\{ \frac{|S_{21\text{AUT}}|^2 \left(1 - |S_{22\text{REF}}|^2\right)}{|S_{21\text{REF}}|^2 \left(1 - |S_{22\text{AUT}}|^2\right)} \right\} \cdot \eta_{\text{REF}} \tag{5.6}$$

which means that in order to calculate the radiation efficiency of the AUT, we need to use the radiation efficiency of the reference antenna in addition to the measured S-parameters.

From the above derivation, we can see that the final result has nothing to do with the transmitting antenna and there is no need for calibration; also the chambers loss does not affect the final result.

Using the standard operating procedures of the RC, in that a measured quantity value is formed as an average from multiple stirrer positions, Equation 5.6 should be written as (5.7).

$$\eta_{AUT} = \left\{ \frac{\left\langle |S_{21AUT}|^2 \right\rangle \times \left(1 - |S_{22REF}|^2\right)}{\left\langle |S_{21REF}|^2 \right\rangle \times \left(1 - |S_{22AUT}|^2\right)} \right\} \times \eta_{REF} \tag{5.7}$$

where $\langle \ \rangle$ = average of the scattering parameters taken from many stirrer positions and $| \ |$ = absolute value. In Equation 5.7, the reflection coefficient quantities are not signified as being from an ensemble average of many stirrer positions. In this form they are assumed to be acquired in an anechoic chamber (AC). It is possible to measure the reflection coefficient of antennas accurately using the RC as shown in Ref. [15], in which case the reflection coefficient quantities in Equation 5.7 should be adjusted to reflect that they would be formed as an average from many stirrer positions.

Before we continue, a few words are warranted on the subject of chamber power losses. It is stated previously that the power at the receiving antenna is assumed to be the same for both the REF and AUT. During any measurement, the procedures have to be carefully controlled in order for this to be realised.

This statement has derived links to the quantity of the chamber Q factor as described in Chapter 2. As we have already discussed, any items added into the chamber will introduce a 'loading' to the chamber which will have some consequence on the chamber Q factor. To ensure that the losses in the chamber are the same as possible, and thus the power reaching each respective antenna (in turn) is not unduly affected, implies that the chamber Q factor needs to be held as constant as possible.

Practical procedures to achieve this normally dictate that all antennas/equipment that are likely to undergo any measurements should be present in the chamber at all times to ensure that the chamber Q factor is as constant as possible. If this is not the case, then subtle uncertainties might ensue from measurement to measurement. Ordinarily, these antennas/equipment inside should have a separation distance between them in order not to induce mutual coupling effects – a minimum of $\lambda/2$ distance is normally considered appropriate.

5.2.2 Total Radiation Efficiency

For the quantity of total radiation efficiency, the IEEE standard definition of terms is again referred to [14]:

The ratio of total radiated power to the power incident on the antenna port.

Mathematically, this is a product of the radiation efficiency and mismatch efficiency of an antenna:

$$\eta_{\text{TOTAL}} = \eta_{\text{AUT}} \times \left(1 - |S_{22\,\text{AUT}}|^2\right) \tag{5.8}$$

From (5.8), this time the impedance matching of the antenna is factored in. This means, for example, if an antenna is poorly impedance matched, say only 1% of power is able to be delivered to the antenna, if the antenna radiates all of that 1% of power, then by definition it will be 1% efficient in terms of its total radiation efficiency. Please note that the format of (5.8) again assumes that the reflection coefficient quantities are measured using an AC.

5.3 Definitions: Textile Antennas

The single-band textile antennas have been developed in Refs. [3, 4] for use in the Industrial, Scientific and Medical (ISM) 2.45 GHz band and are based on a Planar Inverted F Antenna (PIFA) topology. They are constructed using a 6-mm thick felt fabric with (respective) relative permittivity and loss tangent parameters of $\left(\varepsilon_r = 1.43, \tan \delta = 0.025\right)$ at 2.45 GHz [4]. The felt fabric substrate is sandwiched between a slotted radiator at the top and a small ground plane underneath, with two different conductive textile materials being subject to investigation [16, 17].

1. Copper textile (plain woven and coated, 0.08 mm thick, $\sigma = 2.5 \times 10^6$ S/m at 2.45 GHz).
2. ShieldIt™ conductive fabric (0.17 mm thick, $\sigma = 1.18 \times 10^5$ S/m at 2.45 GHz).

The profile and dimensions of each respective antenna can be viewed in Figures 5.1 and 5.2.

5.4 Measurement Procedures

The measurement investigation was carried out using the RC at the University of Liverpool. As stated previously, the chamber has dimensions of width = 3.6 m, height = 4 m and length = 5.8 m. For the efficiency deduction, the chamber required an initial reference measurement for calibration purposes. This calibration was performed using a dual ridge horn antenna (Satimo Model: SH2000), the antenna being selected as its unidirectional pattern characteristics was similar to that of the AUT.

Ordinarily however, an omnidirectional reference antenna could be used. As long as the proportion of power going direct to the receiver in the chamber is minimised (low K factor), then the use of a directive or omnidirectional reference should not

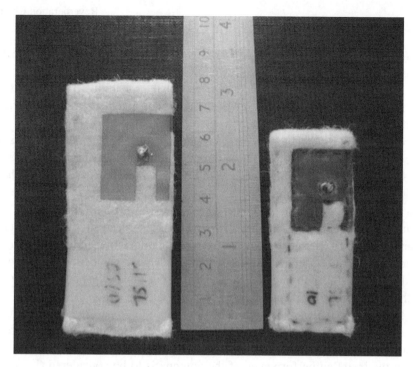

Figure 5.1 Single band textile antenna radiating elements, ShieldIt radiating element (left) and Copper radiating element (right).

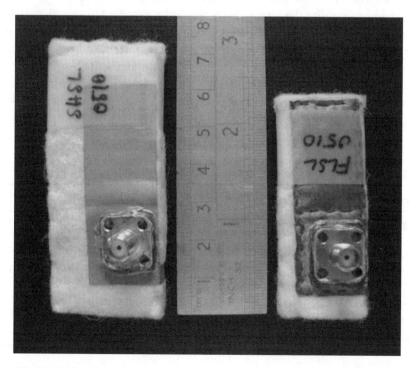

Figure 5.2 Single band textile antenna ground planes, ShieldIt ground plane (left) and Copper ground plane (right).

make too much difference. The reason for this is attributed to the nature of the Angle of Arrival (AoA) of plane waves as shown in Chapter 2 – specifically that they would arrive from every conceivable angle with an equal probability. This characteristic places far less emphasis on the nature of the reference antennas' radiation pattern as a result.

It is crucial therefore that the transmitting antenna during such measurements is not directed towards the receiving antennas directly illuminating them. In our experience it has been found that pointing the transmitting antenna directly at the mechanical stirring paddles minimises the proportion of power travelling direct to the receiver as shown by the K factor results in Chapter 2.

In this work, a frequency range of 2000–3500 MHz is selected for the antenna measurements using 801 frequency data points. The number of data points in a given measurement should be selected to ensure that an adequate number of modes would be excited in the chamber throughout the measurement range which can be calculated from the given theory in Chapter 2.

For free space efficiency measurement, the calibration measurement with the REF antenna needs to be performed with the AUT located inside the chamber for reasons previously explained.

When the installed performance of the antenna is to be considered (i.e. antenna plus human subject), the human 'loading' has to be present in the chamber throughout the calibration phase. This requirement is for the same reason as explained above, and also because the human body would be expected to significantly 'load' the chamber and was surmised to be a dominant contributor to any loss mechanisms that exist inside the chamber (more so than, for example, wall losses, losses from aperture leakage and losses in any antenna). The stirring sequences used for both the free space and on-body measurements, encompassing the reference and AUT measurements in both scenarios, comprised the following parameters as detailed in Table 5.1.

Table 5.1 Measurement parameters for textile antenna efficiency investigation.

Parameter	Description
Stirring sequences	5 degree mechanical stirring
	Polarisation stirring
	5× position stirring
	20 MHz frequency stirring
Total number of measured samples per frequency point	710
Frequencies (MHz)	2000–3500
Number of frequency data points	801
Source power (dBm)	+13

From Table 5.1, the stirring sequences for the measurements have been selected for four primary reasons:

1. A large amount of samples (with a given percentage being sufficiently uncorrelated) are required in order to resolve the efficiency quantity accurately.
2. For measurements using the human subject, it means that the human did not have to spend prolonged amounts of time in the chamber to complete each measured sequence. Hence, the subject could take regular breaks if need be.
3. The exact same number and the manner of the stirring techniques for both free space and on-body measurements means that an accurate comparison could be formed to benchmark performance.
4. The 20 MHz frequency stirring value (averaged over 11 frequency points) was selected such that it was far less than the bandwidth of the antennas to minimise the reduction in frequency resolution (configured to be 4% of the absolute bandwidths of the antennas).

During any measurements in the chamber, it is always a desired practice to terminate any antennas not undergoing measurement with impedance matched loads – this ensures that all antennas inside the chamber are in a balanced condition. For the free space measurements, the AUT was mounted using a dummy cable on a separate platform and terminated in an impedance matched load. This is subsequently transferred to the reference antenna when the AUT is measured.

For on-body reference measurements, the AUT was fixed to the chest of the human subject via the same dummy cable and terminated in the same impedance matched condition. For the on-body AUT measurements, the reference antenna remained on a separate platform in a matched terminated condition. Throughout the on-body measurements, the contents of the human subject's pockets were always emptied (no coins, mobile phone etc.) such that the loading to the chamber would be as constant and consistent as possible. In this, any varying factor that could potentially introduce a source of uncertainty, and that could be removed, was removed to attempt to yield the most consistent and accurate results possible.

The measurement set-up for the on-body reference measurements can be seen in Figure 5.3. This mirrors a typical set-up that is also employed for the free space measurements. For this type, the human loading is obviously removed and the antenna is mounted on a cable which is ordinarily strapped to a non-metallic stand.

The distance between the consecutive position locations marked out in Figure 5.3 was configured such that each location was larger than the correlation distance (about ½ wavelength) to create independent samples – in this, the distances selected were 0.8 m apart. The half wavelength distance is a generic distance that is ordinarily used when platform or position stirring is employed and aims to ensure that progressive locations are uncorrelated from one another, because as we have seen

AUT attached to
human being on
dummy cable

Transmitting
antenna

Reference
antenna

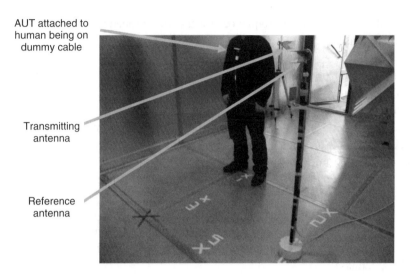

Figure 5.3   On-body reference measurement set-up.

in Chapters 2 and 3, when exact sequences are repeated, no additional stirring is obtained.

Furthermore, it can be seen that a prescribed distance is placed between the human subject and the reference antenna at all times. This is to avoid coupling between the two for reasons previously explained.

5.5   Free Space Measurement Investigation

In this subsection, a discussion will be presented for the free space equivalent values – different on-body locations and proximity distances will be presented in sections to follow. Where appropriate, the measurement results will be issued alongside simulated evidence obtained from CST Microwave Studio (based on the finite integration technique) to depict the accuracy of the measured trends. The simulation efficiency results have been obtained by the gain/directivity method.

The purpose of this section is to visually benchmark the accuracy and appropriateness of the RC results against evidence from modelling tools.

5.5.1   Free Space Performance

With respect to the measurement procedures in Section 5.4 and the measurement parameters in Table 5.1, Figures 5.4, 5.5 and 5.6 depict the free space performance of the copper-based textile (FLSL) and Figures 5.7, 5.8 and 5.9 depict the ShieldIt™ (SHSL) based textile.

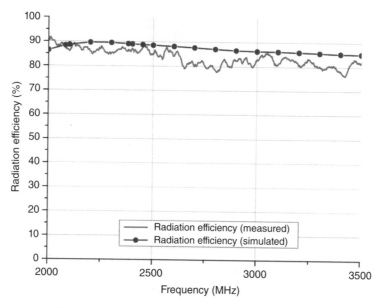

Figure 5.4 FLSL 0510 free space measured and simulated radiation efficiency (%).

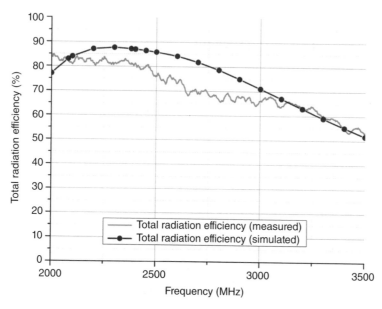

Figure 5.5 FLSL 0510 free space measured and simulated total radiation efficiency (%).

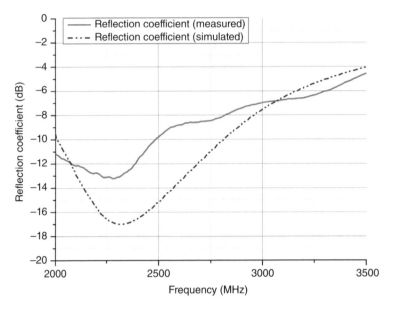

Figure 5.6   FLSL 0510 free space measured and simulated reflection coefficients (dB).

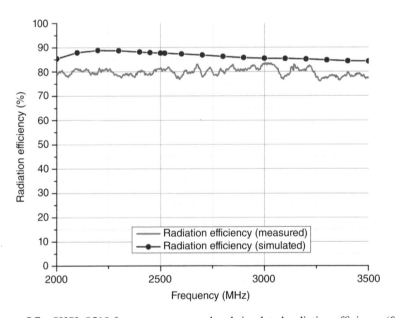

Figure 5.7   SHSL 0510 free space measured and simulated radiation efficiency (%).

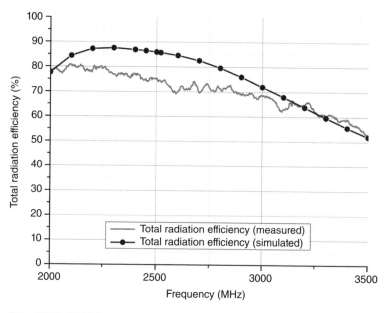

Figure 5.8 SHSL 0510 free space measured and simulated total radiation efficiency (%).

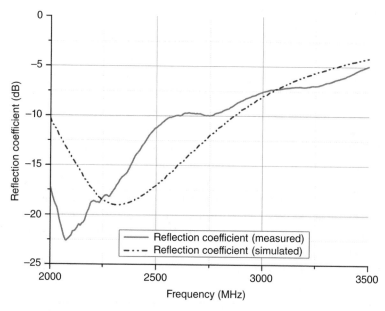

Figure 5.9 SHSL 0510 free space measured and simulated reflection coefficients (dB).

From Figure 5.4 we see that a very good agreement is obtained from both measurement and simulation. In free space, the copper-based textile antenna is seen to be highly efficient.

The efficiency results in this work are chosen to be represented in percentage terms. However, it is possible and different researchers do choose to represent the efficiency quantities in decibel scales. By adopting the use of decibel scales, the measurement uncertainties may begin to provide a different insight. This is because when antennas are highly efficient (say 95% or −0.223 dB), a 5% measurement uncertainty could yield a 'true' value of −0.423 dB, thereby indicating a relatively small difference in decibel terms.

However, when considering antennas presented with a significant loss, a 5% measurement uncertainty in a 20% efficient antenna can yield a significant difference in decibel terms (20% efficiency = −6.987 dB compared to 15% efficiency = −8.239 dB). Hence, however the efficiency data is presented, it is important not only to keep uncertainties in a given measurement to an absolute minimum, but also to understand and properly interpret what the uncertainties mean in terms of the 'true' value that is being sought.

From Figure 5.5 we see a good agreement at the start and end of the frequency range but less so in the mid-range. The discrepancy between the measured and simulated cases can be attributed to differences between the measured and simulated reflection coefficients as shown in Figure 5.6, and not the efficiency measurement.

The discrepancy in the reflection coefficients has resulted from slight differences in the fabricated dimensions between the simulated model and physical design, and also the presence of the SMA connector in the physical design.

Throughout all design stages, care should always be taken in the simulation stage to model the antenna as close to practice as possible and the practical design should be manufactured as close as possible to the desired dimensions as proven in the numerical simulations.

Irrespective of the discrepancy between measured and simulated reflection coefficients, the practical antenna is seen to be sufficiently impedance matched at its desired 2450 MHz centre frequency. Moving on, we can assess the free space radiation efficiency performance of the ShieldIt-based material in Figure 5.7.

From Figure 5.7 we can see relatively high levels of radiation efficiency, which is evident for the ShieldIt-based textile material, but comparing with Figure 5.4 (copper-based textile) they are 8–10% lower, particularly over the first 500 MHz. These results conform completely to expectations, since the higher conductivity material is expected to yield higher levels of radiation efficiency.

Moving on, the total radiation efficiency performance of the ShieldIt-based textile can be viewed in Figure 5.8.

From Figure 5.8 we again see a good agreement between measurement and simulation at the beginning and end of the frequency band, but a discrepancy is also present in the mid-frequency range. This can be explained again by differences between the measured and simulated reflection coefficients, as Figure 5.9 proves.

Nevertheless, from Figure 5.9 we see that the ShieldIt -based textile antenna is sufficiently well impedance matched over the desired band.

In conclusion we find that both antennas are highly efficient and well impedance matched about the frequency band of interest which validates the design premise. In free space we can conclude that the copper-based textile antenna with the higher conductivity exhibits a higher radiation and total radiation efficiency as compared to the lower conductivity ShieldIt material.

The results also show that the RC is a suitable facility to conduct such measurements which is proven by the agreements with the simulation tools. The results also validate the parameters chosen for the measurements. It would be possible to use these parameters as a starting point for measurements in other chambers; however, internal studies should always be performed to assess the measurement accuracy of a chamber to determine the appropriate parameters. This can be performed by use of the uncertainty measurement procedures detailed earlier in Chapter 3.

5.5.2   General Problems to Avoid

At this stage before the measurement discussion progresses further, it is important to discuss some generic problems to avoid when conducting efficiency measurements in the RC.

Please note that this discussion bears no relation to the antennas that are the subject of this chapter, they are generic issues that a given user should be aware of when conducting antenna efficiency measurements.

5.5.2.1   Small Ground Planes

When conducting efficiency measurements on unbalanced antennas with small ground planes, one should always be aware of the nature of surface current flow over the antennas' ground plane. Typically, the surface current can have a tendency to flow easily off a small ground plane structure down towards the cable. If this action occurs, a user at this stage who is conducting efficiency measurements on the device can have little control over this behaviour.

In RC measurements, this effect can manifest itself as a drop in average power that is measured as a result of this current not being transformed into radiated energy by the antenna out into space.

Figure 5.10 depicts this trend and the effect it can have on measured RC results. The results relate to a small-sized unbalanced Ultra Wideband (UWB) antenna and shows that a drop in the average measured power is recorded from the AUT at lower measured frequencies.

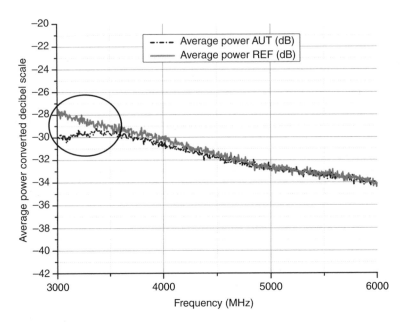

Figure 5.10 Averaged power converted to decibel scale showing 'power drop'.

It should be noted that the antenna was well impedance matched in this range and the drop in measured power is not a result of imperfect impedance matching of the device.

The measured trends when used in Equation 5.7 would predict a lower efficiency in this range as a result of this drop in measured power which ordinarily is not always expected.

To combat such an effect is generally not that straightforward. A Balun could be employed in the measurements as an attempt to block the current from flowing down towards the cable but these tend to be generally narrowband structures.

Another mitigation method could be to modify the antennas' ground plane in the design stage. For UWB antennas, a general technique to minimise the ground plane effects on small-sized antennas was presented in Ref. [18] and advocated a series of cuts on the ground structure to control and confine the surface current.

5.5.2.2 Imperfect Impedance Matching

Although as indicated in Section 5.2.1, the *radiation efficiency* of an antenna can be 100% even if only 1% of input power is accepted by the antenna. At such extreme limits where the antenna is severely mismatched, the measurement of antenna *radiation efficiency* can sometimes become challenging to resolve accurately in the RC.

Depending on the input power levels delivered into the chamber by the transmitting antenna, if these are relatively low then the power levels recorded by a severely mismatched antenna will be still much lower. In some cases, this could begin to approach the 'noise floor' (back ground noise levels) of the chamber which can question the fidelity of the signal being measured. Such low power levels should be avoided.

Generally speaking, the antenna measurement process is usually about proving the merits of a design that has been appropriately designed for operation at a frequency or range of frequencies in which the antenna is sufficiently impedance matched to properly operate; thereby avoiding such an issue.

The antenna does not always need to be impedance matched to levels of −50 dB (reflection coefficient) to be successfully measured in the RC – this is seldom achieved in practice. In our experience, reflection coefficient levels in the order of −0.3 dB magnitude have, for example, proven challenging when attempting to resolve the radiation efficiency accurately, and any given user should at least be aware of this effect.

5.6   On-Body Antenna Measurements

The set up for the on-body measurements has previously been discussed in Section 5.4, and the parameters for this measurement task are detailed in Table 5.1. In this section we are not only going to go through the process of charting how antenna measurements using real human beings can be performed in the RC, but also show how the textile antennas perform when employed in such a role.

It is important at this stage to justify prior to the investigation what proximity distances (antenna/human separation) have been selected. Two distinct proximity distances will be presented; at 0 mm (touching) from the human subject and at 20 mm from the human subject. The purpose of these selections is to depict a kind of 'envelope' that will show the magnitude of the on-body efficiency from strong (expected) body coupling to reduced body coupling to chart the performance with respect to the 2.45 GHz operating frequency. Results will also be presented with the antennas mounted on the chest, back and under loaded (bent) conditions.

For body worn applications when antennas are worn on clothing, movements in the body will cause the distances between the human subject and the antenna to vary, such that a definite separation will not always be evident (i.e. at 5 mm, 10 mm etc.). The distances presented throughout this investigation have been chosen to determine the efficiency at two extreme distances applicable across distinct domains (consumer and emergency services/military applications in this case).

5.6.1   Chest (0 mm) Body Location Investigation

Figure 5.11 portrays the measurement set-up used in this investigation.

Figure 5.11   Chest (0 mm) body worn measurement set-up.

From Figure 5.11, the antenna is located on the centre of the human subjects' chest, 1.38 m from the floor. The 0 mm dimension in this case does not mean that the antenna was physically touching the human subjects' skin; rather, the human subject was wearing a woollen jumper 1.5 mm thick (constant throughout all experiments herewith) and it was this which the antenna was placed against. The antenna element was not physically strapped to the human by any means – standard Velcro straps were used to fix the attached cable to the human being which was sufficient to hold the antenna to the chest of the subject.

Throughout the measurement sequences this aspect was rigorously checked and at no time did the antenna shift location. This aspect is important to note for such experiments as movements in the antenna location could lead to subtle uncertainties being present in any subsequent results.

In Figure 5.11, the human subject faces the back wall of the chamber with the antenna on his chest. The simple aim here is not to face the transmitting antenna which is located behind the human subject (off shot). The antenna orientation in Figure 5.11 (normal to the chest) was chosen because of the location of the SMA connector in the physical design, meaning that it was easier to secure the antenna in position at the 0 mm proximity distance in the manner shown.

The radiation efficiency results for the copper-based textile antenna are disclosed in Figure 5.12. Three separate and independent measurements have been

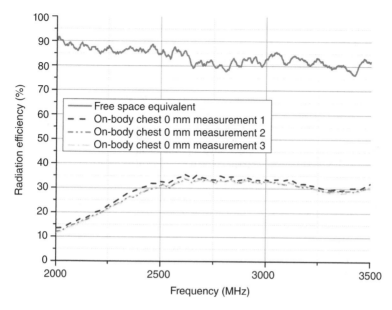

Figure 5.12   Free space and on-body chest (0 mm) radiation efficiency (%).

completed for this antenna in this scenario to provide an indication as to the measurement repeatability on-body. A comparison with the free space measured radiation efficiency is also presented so that the degradation in radiation efficiency can be easily assessed.

From Figure 5.12, it can be seen that the effect of the body at the 0 mm placement distance is severe, and the levels of radiation efficiency are degraded considerably. At the start of the measured band (2000 MHz) this is as much as 78.15%.

The levels of on-body radiation efficiency are also seen to rise, particularly over the first 600 MHz measured band, which is believed to be attributed to reductions in coupling between the human body and the antenna element (remembering the antenna is not physically touching the human skin). Hence, as frequency begins to rise, the inter-element (human body and antenna) spacing is also increasing.

The repeatability in the on-body efficiency measurements using the aforementioned procedures is also interesting to note. A maximum difference of only 2% is found between the three independent measured runs in this case. This result proves that consistent and reliable results can be acquired with the use of human beings if correct and consistent procedures are followed. The on-body total radiation efficiency results for the copper-based textile can be found in Figure 5.13, three independent measured runs are also depicted to chart the measurement repeatability.

Severe deficiencies in performance are again evident. At the start of the measured band there is a 72.65% drop in measured performance. Similar trends as to the

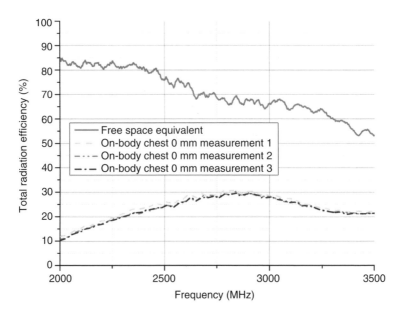

Figure 5.13   Free space and on-body chest (0 mm) total radiation efficiency (%).

radiation efficiency are witnessed and the repeatability levels are still in the order of 2% between each of the three measured runs undertaken.

Questions could arise with respect to the antennas on-body orientation (from Figure 5.11), particularly as the antenna is mounted normal to the chest instead of being parallel; an investigation was undertaken with a different on-body orientation. The purpose for this investigation is because in a worn application, it is well known that the antenna orientation can change due to the users' movements. Therefore, even when a wearable antenna is designed specifically to be radiating along the body or away from the body, neither of these orientations will exist consistently at all times.

Therefore, a need exists for two distinct evaluations, one radiating along the body (normal to the chest) and one radiating away from the body (parallel to the chest). Moreover, fixing the antenna in an on-body role beats the purpose of having a wearable antenna which can offer the flexibility and comfort that conventional- (metallic) type antennas cannot.

The measurement set-up for the different orientation measurement can be viewed in Figure 5.14.

This investigation utilised the exact same measurement parameters and procedures as previously discussed to ensure that an accurate comparison could be formed. The on-body antenna location was configured such that the cable was mounted and secured under the arm, enabling the antenna radiating patch to be away from the human body. The exact same (0 mm) proximity distance was also maintained. Figure 5.15 illustrates

Figure 5.14 (0 mm) Measurement set-up for alternative orientation.

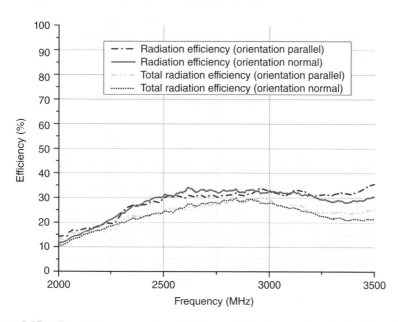

Figure 5.15 (0 mm) Measurement results for parallel and normal on-body orientations.

the measurement results from this investigation and the results are directly compared alongside the previous normal orientation.

The results in Figure 5.15 show that irrespective of the antennas on-body orientation, the deduced radiation and total radiation efficiency results are very close indeed; the difference at the largest point is in the order of 4%. The result proves that in this particular case it does not matter too much what the orientation is (i.e. the results are valid irrespective of the orientation). The question now is: Why is this so?

- In a Line of Sight (LoS) environment, the on-body orientation will be crucial because of the radiation pattern characteristics with respect to how the antenna is orientated on-body. In the LoS scenario, such an agreement (as in Figure 5.15) is not necessarily expected. In the Non-Line of Sight (NLoS) environment however, as proven back in Chapter 2, there is a uniform distribution of plane wave AoAs, meaning that the probability of the wave coming from every conceivable direction will be equal. The existence of this environment can therefore serve to mitigate the effect of the antennas' radiation pattern with respect to the on-body orientation. A similar argument to this was put forward in Ref. [16].
- Another contributing factor concerns the size of the ground plane. If a large ground plane size was to exist, then such an agreement is again not expected – that is, a more pronounced difference is envisaged as the on-body antenna pattern characteristics would be more strongly influenced by the larger ground plane.

The parallel measured results from Figure 5.15 were acquired on a different human subject when compared to the normal measured results. It can be concluded therefore that any results measured at the 0 mm proximity distance are comparable, irrespective of the on-body antennas' orientation (within reason), and second that the measured results are reproducible and representative of different human subjects. From herewith, all results measured at the 0 mm proximity distance have used the normal antenna orientation (as in Figure 5.11) due to the ease of this set-up.

As previously stated, all measured reflection coefficients have been acquired using an AC. The on-body locations for the RC measurements were measured and marked, such that for the AC measurements, the antennas were placed in exactly the same position to ensure that accuracy ensued. Figure 5.16 details the typical measurement set-up and Figure 5.17 depicts the on-body reflection coefficient performance.

The set-up in Figure 5.16 shows that the exact same Velcro straps have been employed and the antenna is fixed in the exact same manner as the RC measurements. This aspect is important in order to ensure consistency between the body locations that have been chosen and the respective results that have been acquired.

From Figure 5.17, in addition to the severely degraded radiation efficiency performance, the antenna is also seen to exhibit a degree of de-tuning. At 2.45 GHz the mismatch efficiency measured on the chest is 78% as compared to the free space value of 91%.

Figure 5.16 (0 mm) Reflection coefficient measurement set-up.

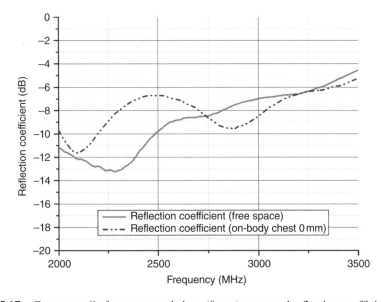

Figure 5.17 Copper textile free space and chest (0 mm) measured reflection coefficients (dB).

We can now assess the performance of the lower conductivity material textile antenna (ShieldIt) and compare it against the higher conductivity copper- based textile antenna to see if any marked difference in performance is evident.

Figures 5.18 and 5.19 depict the on-body efficiency and reflection coefficient performance, respectively for the ShieldIt material textile antenna.

If we compare Figures 5.12 and 5.13 with Figure 5.18 we find that the antenna constructed with the ShieldIt-type material is lossier in free space conditions than the higher conductivity copper-based textile. As previously established, this is completely in line with expectations.

However, when the on-body efficiencies are compared, we find that the lower conductivity (lossier free space) antenna is grossly more efficient than its copper-based counterpart. The difference between the performances of the two textile antennas on-body (chest 0 mm) yields up to 30% for the radiation efficiency and 25% for the total radiation efficiency.

Therefore, when assessing the efficiency degradation on-body, the ShieldIt material, in this case, has by far the superior performance. An assessment of the reflection coefficient performance on-body for ShieldIt textile antenna is shown in Figure 5.19.

From Figure 5.19 it can be seen that evidence of de-tuning also exists but it is not as severe as for the higher conductivity (copper-based) textile antenna. Comparing the mismatch efficiencies of the two textile antennas at 0 mm, we find that the ShieldIt textile material exhibits a value of 90% as opposed to the 78% witnessed from the

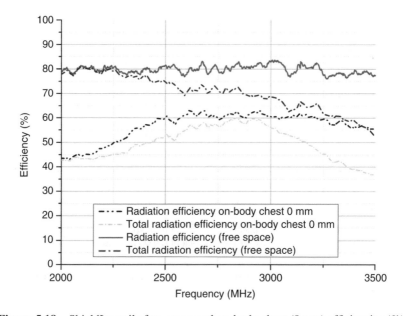

Figure 5.18   ShieldIt textile free space and on-body chest (0 mm) efficiencies (%).

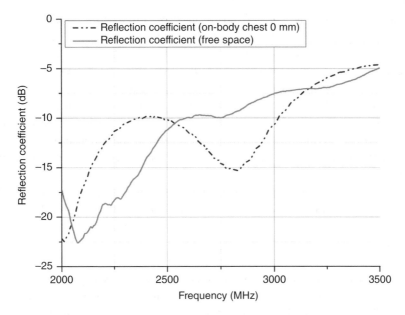

Figure 5.19   ShieldIt textile free space and chest (0 mm) measured reflection coefficients (dB).

copper textile material. About the designed centre frequency, the ShieldIt textile antenna remains well matched.

The results so far suggest that the lower conductivity material is a more suitable choice for operation on-body as the detriment in the efficiency and de-tuning performance is lower.

5.6.2   Elbow (0 mm) Bent Location Investigation

This investigation looks to assess the antennas' performance under a bent condition. The purpose of this investigation is to demonstrate what can happen to the antennas' performance if it becomes stressed and unnecessarily flexed due to human movements.

The on-body location area used to perform this experiment was chosen to be the elbow. This location was selected as opposed to the knee for example, because the antenna could not be located so close to a metallic boundary (i.e. the floor) for field uniformity purposes as detailed in Chapter 2.

In this experiment, the antenna has to be secured with the standard Velcro ties owing to the loading placed on the antenna through bending and to keep the antenna element fixed. The bending radius (measured from the centre of the elbow to the edge of the elbow) was configured at 55 mm. The antenna element selected to perform this test was the copper-based textile and was located 1.2 m from the chamber floor.

The human subjects' arm was fixed in place by a Velcro tie which was strapped around the arm and the shoulder; this was sufficient to hold the arm in place and prevent any unnecessary movement. The RC measurement set-up can be seen in Figure 5.20 and the radiation efficiency, total radiation efficiency and reflection coefficient performance in this measurement scenario can be seen in Figure 5.21.

Comparing the bent results with the chest (0 mm) measurement results from Figure 5.12, the radiation efficiency results at 2 GHz would appear to be slightly higher in the bent configuration; this is believed to be due to the absence of major human organs near the antenna. Throughout the mid- and upper-frequency range the radiation efficiency values are comparable.

However, the main effect here concerns the total radiation efficiency. The levels are 12–15% lower than the 0 mm chest values from Figure 5.13 owing to the severe detuning in the magnitude of the reflection coefficient; rendering the mismatch efficiency in the bent configuration at 62% as compared to 78% on the chest at 0 mm and 91% in free space.

It is worth acknowledging at this stage that the antenna was stressed for a considerable period of time, more so than, for example, if it was employed in a real on-body scenario. However, in a real scenario, if the antenna was in a location on the body that was liable to bend, then at any instantaneous point in time the antenna could severely

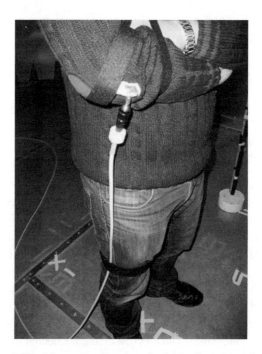

Figure 5.20   Measurement set-up for bent elbow investigation.

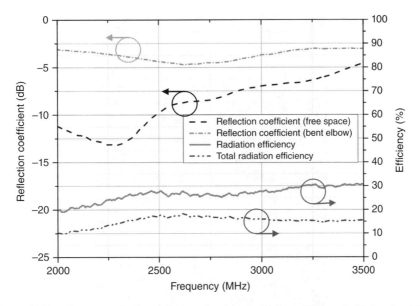

Figure 5.21   Copper textile bent elbow reflection coefficients (dB) and efficiencies (%).

de-tune and this could have consequences on the communication link. The conclusion drawn from this investigation is that severely loading the antenna under a bent configuration is not a good practice, and on the evidence of the results seen in Figure 5.21, should be avoided if possible when an on-body location is to be chosen for the antennas' placement.

5.6.3   Back (0 mm) Body Location Investigation

This investigation looks to assess if any fundamental differences in performance are witnessed at different on-body locations. Any differences, if revealed, would not necessarily be a result of differing pattern characteristics with respect to different on-body locations within the RC environment; rather, a difference in coupling levels between the human body and the antenna as a function of body location.

The option selected for this investigation was the human subjects' back. The reason for this selection was to situate the antenna away from the main human organs and not in a scenario where it was likely bent severely. The RC measurement set-up can be seen in Figure 5.22, the efficiency and reflection coefficient results for copper textile antenna can be found in Figure 5.23. The results for the ShieldIt material textile are presented in Figure 5.24. The exact same measurement procedures previously discussed have also been adopted in this investigation.

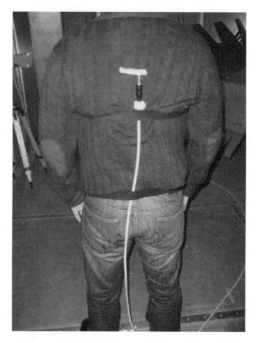

Figure 5.22   Measurement set-up for back investigation.

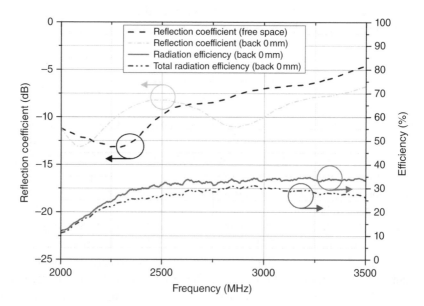

Figure 5.23   Copper textile back (0 mm) reflection coefficients (dB) and efficiencies (%).

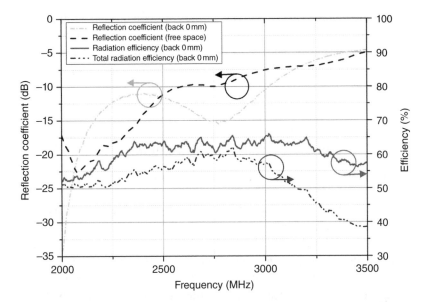

Figure 5.24   ShieldIt back (0 mm) reflection coefficients (dB) and efficiencies (%).

Comparing the radiation efficiency performance of the copper textile on the back at 0 mm to that of the chest at 0 mm (Figure 5.12) it is found that the results on the back are generally comparable in nature. There is an increase of 5% towards the end of the measured band however. The results for the total radiation efficiency are generally comparable in nature with the chest (0 mm) values, although they are slightly higher by a few percent.

There is also frequency de-tuning evident with the antenna at this on-body location. At the centre frequency, the mismatch efficiency on the back is recorded as 85% as opposed to the 78% witnessed on the chest and 91% measured under free space conditions. From the evidence witnessed it can be concluded that no wholesale differences in performance are apparent from the change in on-body location.

While assessing the ShieldIt material on the back at 0 mm, please note the slight shift in scale on the right-hand y axis. The results of these are presented in Figure 5.24.

From Figure 5.24 it can be seen that again the lower conductivity ShieldIt- based material has superior performance on-body as compared to the higher conductivity copper textile. The radiation efficiency and total radiation efficiency results for this antenna on the back are again comparable with the levels witnessed on the chest (at 0 mm).

Comparing the magnitude of the reflection coefficient obtained on the chest (0 mm) and in free space, we find that on the back no negative de-tuning is evident whatsoever in this on-body location.

The results in Figure 5.24 re-enforce the results obtained in Figure 5.18 (chest 0 mm), in so much that any measurement anomaly or error can be ruled out – in different on-body scenarios the lossier free space antenna continues to outperform the higher free space antenna (copper textile), both in terms of efficiency and reflection coefficient performance.

5.6.4   Chest 20 mm Body Location Investigation

The previous investigations have established the following:

1. A lossier antenna in free space grossly outperforms a higher free space efficient antenna when placed on-body.
2. The radiation efficiency values obtained in the RC environment for each respective antenna are within a few percent of the values obtained at different on-body locations.
3. The total radiation efficiency values are subject to variations at different on-body locations due to the added dependence of this quantity with the on-body reflection coefficient performance of the antenna.

The next investigation looks to assess the effect of having different spacing between the antenna and the human body – that is if the inter-element spacing were to be slightly increased, what difference in the magnitude of radiation and total radiation efficiency could be obtained?

The theoretical rationale guiding this investigation is considered to be a function of (expected) decreased coupling between the antenna and the human subject; noting that the medium that is coupling to the antenna (i.e. the human being) will present a loss (sometimes significant) if the antenna is close enough.

As previously stated, the 20 mm proximity distance was chosen to complete the 'envelope' from strong expected coupling (very close to the body) to reduced coupling (slightly away from the body), maintaining a distance that can still be applicable in an on-body scenario. The practical application for such a 20 mm distance from the body can be realised for emergency service workers who may have antennas attached to their overcoats.

In this investigation the antenna was placed 1.38 m from the floor and situated from the human subjects' chest. To achieve and maintain the 20 mm prescribed distance from the body, the antenna was attached via a 90° 'elbow' connector. Figure 5.25 details the measurement set-up for this investigation.

From Figure 5.25 it can be seen that with the presence of the elbow connector, the antenna is orientated parallel with the body. Consistent with the previous on-body investigations, Velcro straps have again been used to fix the cable to the human subject to prevent any movements, and the exact same measurement procedures have again been employed.

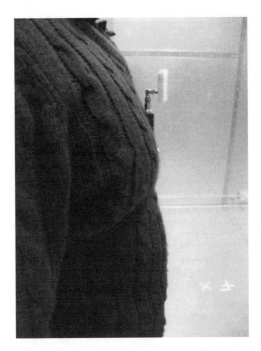

Figure 5.25   Measurement set-up for the chest (20 mm) investigation.

Figure 5.26 details the on-body efficiency and reflection coefficient performance of the copper textile antenna.

Comparing Figures 5.12, 5.23 and 5.26, that is on the chest at 0 mm, on the back at 0 mm and on the chest at 20 mm, it can be clearly seen that the radiation and total radiation efficiency values are at a higher level at 20 mm than previously deduced. A 20–30% increase in the radiation efficiency is apparent in this case and a 15–19% increase in the total radiation efficiency. Furthermore, the radiation efficiency levels in Figure 5.26 are also seen to rise with increasing frequency which provides confidence to believe that the coupling theory can explain this trend.

It is also revealed that the added distance from the human subject has reduced the loading effects on the antenna sufficiently enough such that no negative de-tuning is apparent in the on-body reflection coefficient. The performance of ShieldIt textile at the 20 mm proximity distance can be viewed in Figure 5.27. Please note again the slight shift in the right-hand y axis.

Comparing Figures 5.18, 5.24 and 5.27, that is, on the chest at 0 mm, on the back at 0 mm and on the chest at 20 mm, it is again evident that the added proximity distance mitigates the detrimental effect of the human body. This trend can be explained again by the reduced coupling between the antenna and human body due to the added (electrical) distance between both elements. Further, by assessing the magnitude of the

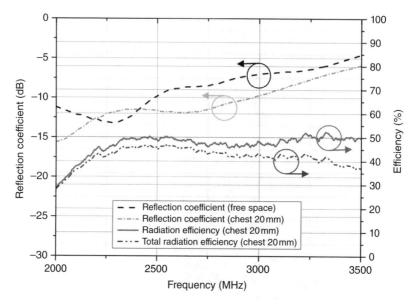

Figure 5.26   Copper textile chest (20 mm) reflection coefficients (dB) and efficiencies (%).

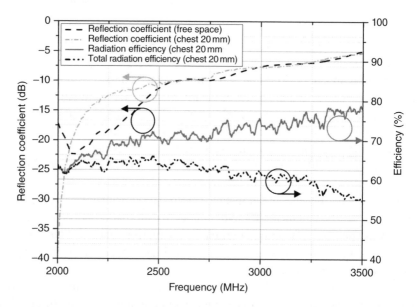

Figure 5.27   SHSL 0510 chest (20 mm) reflection coefficients (dB) and efficiencies (%).

radiation efficiency at 20 mm, it can be seen that the antenna towards the end of the measured range has decoupled itself from the human body to such an extent that the on-body radiation efficiency levels approach the levels evident in free space.

This fact could prove useful when antennas with multi-band operational frequencies are used (progressively higher frequencies), such that the higher frequency bands suffer less of a radiation efficiency degradation due to antenna/body coupling – particularly so when located 20 mm off the body as seen in this work.

If we assess the effect on the measured reflection coefficient, we find that again no negative de-tuning has taken place at all at the 20 mm distance using this antenna – the antenna is operational over a similar bandwidth as free space. It is proven therefore that certain antennas can be placed in (relatively) close proximity to the human body in an operational role and suffer only minor detrimental performance effects.

5.7 Theoretical and Simulated Evidence

Drawing conclusions from Section 5.6, it is clearly proven that the antenna constructed from the lower conductivity, thicker textile material (ShieldIt) outperforms the thinner, higher conductivity copper textile antenna when placed in various locations and proximity distances of the human body, in terms of both the efficiency and the frequency de-tuning levels. This result is in stark contrast to the free space (efficiency) case and counter-intuitive to what we would normally expect.

The question therefore is: Why is this the case?

It is stated in Ref. [17] that demands placed upon antennas with small ground planes, when placed near the proximity of a human body, can result in an interaction with the reactive near fields of the antenna and cause a loss. The reason here for the difference in the magnitude of efficiency between the two antennas is due to the fact that the lower conductivity (thicker) textile material has given rise to lower electric fields in the body as more power has been lost in this antenna itself for the same input power. This fact is understood to have had the effect of causing lower losses in the human body as opposed to the higher conductivity (copper) textile material [19].

To help reinforce this statement and provide evidence that this theory can explain the trends, two simulated models have been adopted using a well-known commercial EM solver (CST Microwave Studio). The models have been mounted (at 1.5 mm to reflect the woollen jumper worn in practice) onto a structure whose material parameters have been chosen to emulate muscle at 2.45 GHz ($\sigma = 1.773$ S/m and $\varepsilon_r = 52.668$ [20]). Figure 5.28 depicts the simulated radiated power and radiation efficiency from the two models, showing clearly that the copper-based textile antenna radiates less power into free space than the SHSL counterpart. Further, by definition, it is also seen to be less efficient when placed in conjunction with the simulated lossy structure. Figure 5.29 depicts the simulated electric field

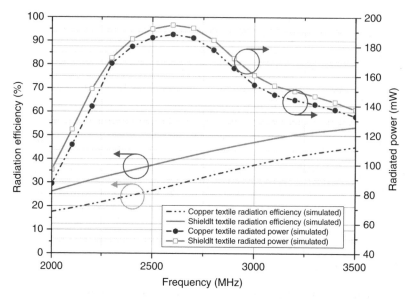

Figure 5.28   Simulated radiation efficiency and radiated power on muscle emulated material.

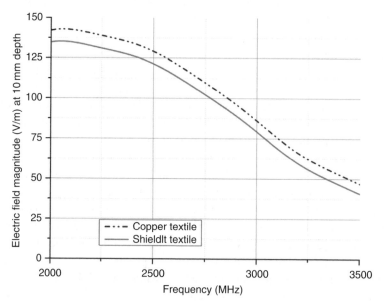

Figure 5.29   Simulated electric field magnitude at 10 mm depth inside muscle emulated material.

magnitudes inside the emulated muscle structure at a depth of 10 mm. A clear difference is shown between the two antenna models which reinforces the theory that explains this trend.

5.8 Measurement Uncertainty

For the measurement uncertainties the following aspects are considered. First, it is known from Ref. [21] that direct coupling can be a major source of uncertainty during Over-the-Air (OTA) measurements in the RC; therefore, the direct coupling (expressed as the Rican K factor as detailed in Chapter 2) should be as small as possible.

Models were presented in Ref. [21] that equate the total standard deviation in the average power transfer function to be comprised of the Non-Line of Sight (NLoS) number of independent samples (σ_{NLOS}), the Line of Sight Number (LoS) of independent samples (σ_{LOS}) and the Rician K factor as detailed in Equation 5.9.

$$\sigma = \sqrt{\frac{\left(\sigma_{\mathrm{NLOS}}\right)^2 + \left(K_{\mathrm{av}}\right)^2 \left(\sigma_{\mathrm{LOS}}\right)^2}{\sqrt{1 + \left(K_{\mathrm{av}}\right)^2}}} \tag{5.9}$$

where (K_{av}) is the average Rician K factor, comprising the samples obtained from the different stirring mechanisms used throughout the measurements as detailed in Table 5.1. The standard deviation in decibel scale can be presented as an average of $(1+\sigma)$ and $(1-\sigma)$ [22], and is detailed in (5.10).

$$\sigma_{\mathrm{dB}} = \frac{10\{\log_{10}(1+\sigma) - \log_{10}(1-\sigma)\}}{2} = 5\log_{10}\left(\frac{(1+\sigma)}{(1-\sigma)}\right) \tag{5.10}$$

To calculate the NLoS number of independent samples, the autocorrelation function was employed as defined in Ref. [23]. After which, [24] was referred to and repeated in terms of the total number of measurement samples used throughout this investigation (710 samples per frequency point), to obtain the correct critical value for use in the autocorrelation calculation at a 99% confidence interval; thus the $1/e = 0.37$ criterion *was not used* in this investigation.

The full procedure of how the NLoS number of independent samples is calculated is detailed in Appendix A.

The LoS number of independent samples can be calculated by (5.11) after consultation with [19].

$$\sigma_{\mathrm{LOS}} = N_{\mathrm{PL}} \times N_{\mathrm{ANTENNA\_IND}} \tag{5.11}$$

where N_{PL} = number of position locations = 5 and $N_{\mathrm{ANTENNA\_IND}}$ = number of independent transmitting antenna locations = 1.

Before any standard deviations are issued, it is important to assess the statistics of the measurement on the different on-body locations used throughout the investigation. Any difference in the measured statistics at different on-body locations could result in a different uncertainty, which is unacceptable. Figure 5.30 details the measured Rician K factor at different on-body locations for the single band textile antennas. A result taken from the free space textile antenna measurements (i.e. no human presence) is also provided to depict the magnitude of the human 'loading' to the chamber.

Figure 5.30 shows that a slight increase in the proportion of direct power is evident in the on-body measurements from the free space equivalent (no human presence), but this increase is not seen to be large as to expect a significant rise in the measurement uncertainty resulting from this quantity.

Figure 5.30 also proves that the measured statistical environment is consistent from one body location to another. No major fluctuations are present which means that similar uncertainty levels are to be expected irrespective of different on-body locations. With respect to different human subjects, Figure 5.31 depicts the effect on the proportion of direct power in the chamber. Two male and one female subjects are employed with heights ranging from 1.74 to 1.8 m and weights ranging from 70.5 to 81.3 kg, each subject wearing different clothing.

Figure 5.31 shows that the statistics are comparable, irrespective of the human being, proving that a consistent platform in the chamber is realised from one human being to the next.

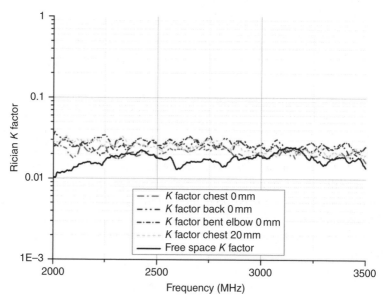

Figure 5.30 Measured Rician K factor in free space and at different on-body locations for single band antennas.

The final standard deviations can now be issued. Figure 5.32 depicts the standard deviations in linear and decibel scale for the single band antennas as a function of different on-body locations, where it can be seen that a very low overall uncertainty exists (in the order of 0.22 dB for all the different locations tested). The uncertainty is also

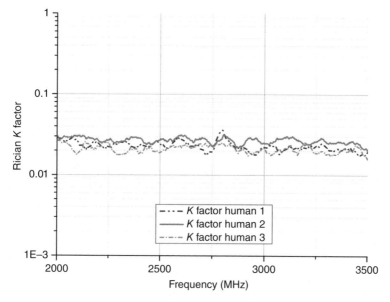

Figure 5.31 Measured Rician *K* factor for different human subjects.

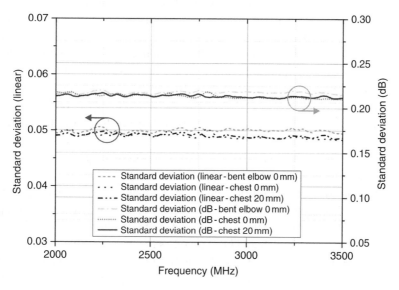

Figure 5.32 Measurement uncertainties (linear and decibel) for single band investigation.

seen to be comparable, irrespective of the on-body location chosen, which proves that any different on-body locations or any slight movements that may have occurred in the measurements do not affect the measurement repeatability and/or accuracy.

Please note that the scales in Figure 5.32 do not represent a direct conversion from linear to decibel and vice-versa. The graphs have been scaled such that both information sets are clearly visible.

5.9   Summary

This chapter has shown how to use an RC to measure the radiation efficiency of single-port antennas. As examples, efficiency measurements performed using textile antennas on live human beings have been conducted in an accurate and controlled manner, with the repeatability being as close as 2%. This confirms that any subtle movements by the human being have not contributed to any serious deviations in the measurement. The magnitude of on-body losses experienced by a given textile antenna with a small ground plane is seen in this case to be a function of the material properties of that antenna – a higher conductivity, thinner (copper-based) textile material which is more efficient in free space is seen to perform worse when placed on- body as compared to a lower conductivity and lossier free space textile antenna with the same overall design topology.

This is proven to be due to the fact that the lower conductivity material- based antenna has given rise to lower electric fields in the body as more power has been lost in the antenna itself for the same input power. This is a remarkable result, constituting new knowledge which can have a profound impact on the material choice for these small-sized antennas, in the sense that a higher conductivity material would appear not always to be the best option when operating in close proximity to a human being.

The magnitude of efficiency losses on body has been experimentally shown to be mitigated somewhat by a variation in the distance from the body – a small 20 mm distance from the body (for antenna SHSL) in this case was sufficient to show that a reduction in radiation efficiency can be eliminated by up to 22%. For the ShieldIt textile antenna, at higher frequencies the 20 mm (off) body result approached the radiation efficiency value in free space.

From the experiment that looked to assess the effect of loading the antenna via bending, one can conclude that this aspect is not a good practice – the antenna was severely loaded for a considerable period of time (more so than, for example, if placed on a body part that relaxed the antenna from time to time), and thus the de-tuning performance was seen to be severe in this scenario. From the results seen here this condition cannot be recommended and should be avoided as much as possible.

On the whole, the investigation has confirmed that the textile antenna is a strong candidate for use in the on-body role. With regards to the measurement practices and facility, the methods utilised have yielded an accurate and repeatable series of results;

this validates the RC's use for this type of measurement. It has also been proven that the statistical measurement environment is consistent, irrespective of different on-body locations or different human subjects. This is a useful fact, proving the suitability of the RC facility.

References

[1] P. Salonen, L. Sydanheimo, M. Keskilammi and M. Kivikoski, 'A small planar inverted-F antenna for wearable applications', *The Third International Symposium on in Wearable Computers, 1999. Digest of Papers*, IEEE, 18–19 October 1999, Francisco, CA, pp. 95–100.

[2] P. Salonen, M. Keskilammi, J. Rantanen and L. Sydanheimo, 'A novel Bluetooth antenna on flexible substrate for smart clothing', *2001 IEEE International Conference on Systems, Man, and Cybernetics*, vol.2, IEEE, 2001, pp. 789–794.

[3] P. J. Soh, G. A. E. Vandenbosch, S. L. Ooi and M. R. N. Husna, 'Wearable dual-band Sierpinski fractal PIFA using conductive fabric', *Electronics Letters*, vol. 47, p. 365, 2011.

[4] P. J. Soh, G. A. E. Vandenbosch, O. Soo Liam and N. H. M. Rais, 'Design of a broadband all-textile slotted PIFA', *IEEE Transactions on Antennas and Propagation*, vol. 60, pp. 379–384, 2012.

[5] M. Klemm and G. Troester, 'Textile UWB antennas for wireless body area networks', *IEEE Transactions on Antennas and Propagation*, vol. 54, pp. 3192–3197, 2006.

[6] W. Zheyu, Z. Lanlin, D. Psychoudakis and J. L. Volakis, 'GSM and Wi-Fi textile antenna for high data rate communications', *2012 IEEE in Antennas and Propagation Society International Symposium (APSURSI)*, July 2012, Chicago, IL, pp. 1–2.

[7] J. Lilja, P. Salonen, T. Kaija and P. De Maagt, 'Design and manufacturing of robust textile antennas for harsh environments', *IEEE Transactions on Antennas and Propagation*, vol. 60, pp. 4130–4140, 2012.

[8] M. Hirvonen, C. Bohme, D. Severac and M. Maman, 'On-body propagation performance with textile antennas at 867 MHz', *IEEE Transactions on Antennas and Propagation*, vol. 61, pp. 2195–2199, 2013.

[9] Q. Bai and R. J. Langley, 'Effect of bending and crumpling on textile antennas', *2009 2nd IET Seminar on in Antennas and Propagation for Body-Centric Wireless Communications*, 20 April 2009, London, pp. 1–4.

[10] Z. H. Hu, Y. I. Nechayev, P. S. Hall, C. C. Constantinou and H. Yang, 'Measurements and statistical analysis of on-body channel Fading at 2.45 GHz', *IEEE Antennas and Wireless Propagation Letters*, vol. 6, pp. 612–615, 2007.

[11] D. B. Smith, L. W. Hanlen, J. Zhang, D. Miniutti, D. Rodda and B. Gilbert, 'First- and second-order statistical characterizations of the dynamic body area propagation channel of various bandwidths', *Annals of Telecommunications*, vol. 66, pp. 187–203, 2010.

[12] K. Minseok and J. I. Takada, 'Statistical model for 4.5 GHz narrowband on-body propagation channel with specific actions', *IEEE Antennas and Wireless Propagation Letters*, vol. 8, pp. 1250–1254, 2009.

[13] G. A. Conway, W. G. Scanlon, C. Orlenius and C. Walker, 'In situ measurement of UHF wearable antenna radiation efficiency using a reverberation chamber', *IEEE Antennas and Wireless Propagation Letters*, vol. 7, pp. 271–274, 2008.

[14] 'IEEE Standard Definitions of Terms for Antennas', *IEEE Std 145-1993*, p. i, 1993.

[15] X. Chen, P. S. Kildal, 'Accuracy of antenna input reflection coefficient and mismatch factor measured in reverberation chamber', *Third European Conference on Antennas and Propagation, 2009. EuCAP 2009*, 23–27 March 2009, Berlin, pp. 2678–2681.

[16] W. G. Scanlon and N. E. Evans, 'Numerical analysis of bodyworn UHF antenna systems', *Electronics & Communication Engineering Journal*, vol. 13, pp. 53–64, 2001.

[17] P. S. Hall, H. Yang, Y. I. Nechayev, A. Alomalny, C. C. Constantinou, C. Parini, M. Kamarudin, T. Salim, D. Hee, R. Dubrovka, A. Owadally, W. Song, A. Serra, P. Nepa, M. Gallo, M. Bozzetti, 'Antennas and propagation for on-body communication systems', *IEEE Antennas and Propagation Magazine*, vol. 49, pp. 41–58, 2007.

[18] Y. Lu, Y. Huang, H. T. Chattha, Y. C. Shen and S. J. Boyes, 'An elliptical UWB monopole antenna with reduced ground plane effects', *2010 International Workshop on Antenna Technology (iWAT)*, IEEE, 1–3 March 2010, Lisbon, pp. 1–4.

[19] S. J. Boyes, P. J. Soh, Y. Huang, G. A. E. Vandenbosch and N. Khiabani, 'Measurement and performance of textile antenna efficiency on a human body in a reverberation chamber', *IEEE Transactions on Antennas and Propagation,* vol. 61, no. 2, 871–881, 2012.

[20] P. S. Hall and Y. Hao, *Antennas and Propagation for Body-centric Wireless Communications*, 2nd ed.: Boston, MA: Artech House, 2012.

[21] P. S. Kildal, S. H. Lai and X. M. Chen, 'Direct coupling as a residual error contribution during OTA measurements of wireless devices in reverberation chamber', *2009 IEEE Antennas and Propagation Society International Symposium and Usnc/Ursi National Radio Science Meeting*, vols 1–6, IEEE, June 2009, pp. 1428–1431.

[22] P. S. Kildal, X. Chen, C. Orlenius, M. Franzen and C. S. L. Patane, 'Characterization of reverberation chambers for OTA measurements of wireless devices: Physical formulations of channel matrix and new uncertainty formula', *IEEE Transactions on Antennas and Propagation*, vol. 60, pp. 3875–3891, 2012.

[23] BS EN 61000-4-21:2011 'Electromagnetic compatibility (EMC): Testing and measurement techniques. Reverberation chamber test methods', ed., 2011.

[24] H. G. Krauthauser, T. Winzerling, J. Nitsch, N. Eulig and A. Enders, 'Statistical interpretation of autocorrelation coefficients for fields in mode-stiffed chambers', *EMC 2005: IEEE International Symposium on Electromagnetic Compatibility*, vols 1–3, pp. 550–555, 2005.

6

Multiport and Array Antennas

In this chapter, we will see how the Reverberation Chamber (RC) can be used to assess the performance of multi-port antennas. A distinction will be offered between antennas for Multiple Input Multiple Output (MIMO) applications and more conventional array-type antennas. The concepts of diversity gain, embedded element efficiency and channel capacity will then be introduced and followed by a discussion on a technique to measure the radiation efficiency of conventional array antennas in the RC; this will include the development and use of a new characteristic equation. Throughout, all measurement procedures and equations will be documented, as well as a discussion on subtle problems to watch out for during practical measurement tasks.

6.1 Introduction

Antenna arrays have many different performance parameters which can be applied as an indicative measure of the merits in a given design. Traditionally, the performance of such arrays has always been characterised in the anechoic chamber (AC) or free space in conjunction with the much sought after radiation pattern characteristics of a given device. However, over the past few years the RC has emerged as a promising candidate for the measurement of multi-port antenna parameters [1, 2]. In these prior published works, the measurement techniques have considered the 'embedded element' – that is, one element excited, while all other elements are terminated in impedance matched loads.

Reverberation Chambers: Theory and Applications to EMC and Antenna Measurements, First Edition.
Stephen J. Boyes and Yi Huang.
© 2016 John Wiley & Sons, Ltd. Published 2016 by John Wiley & Sons, Ltd.

For MIMO antenna characterisation, the embedded element approach is desired because access is required to characterise each individual measured 'channel' in turn. The essential requirement for MIMO antennas is that diverse reception must be provided – that is, the different ports on one single device must be capable of receiving different signals; which implies that the signals at different ports must be sufficiently uncorrelated from one another. This way, when employed in a MIMO system (which consists of several transmitters and receivers), the antenna can provide for a high data rate/high capacity. Furthermore, when the different signals are combined according to given diversity techniques, they are able to combat fading in multi-path propagation environment [3, 4].

Antenna diversity can be realised in various ways. Depending on the environment and the interference that is expected, designers usually have a choice of which scheme to apply. The most common techniques include spatial, polarisation and pattern diversity. Spatial diversity is performed with multiple antennas, usually having the same characteristics, which are physically separated by space. Depending on the environment, sometimes a distance in the order of half wavelength is sufficient [2].

Spatial diversity can become impractical however when designing multiple antennas on small devices such as mobile telephones and laptops because the overall space available is limited. The addition of multiple feeds in close proximity to one another will inevitably result in high mutual coupling and increased correlation so other techniques are usually required.

For multi-port arrays that are not designed for MIMO applications, a distinction should perhaps be created, as in their practical operation they are typically used in an 'all excited' manner. Typically, this 'all excited' scenario could be active in nature; that is, an array of elements excited individually by separate generators [5]. However, this active nature could be complicated to analyse and would deter from the convenient nature of measurements in the RC that use standard network analysers. Thus, from a measurement perspective, a concession can be made to treat the analysis of multi-port arrays via a 'passive' approach instead; that is, a single excitation source exciting all the array elements through a series of power dividers [5]. This would retain the favourable nature of using standard network analysers but would begin to encompass the 'all excited', realistic nature of the array operation.

Another benefit to this approach is that the radiation efficiency measurement of the entire array; important because the radiation efficiency performance of such arrays is liable to be affected by mutual coupling, can be effectively treated in a manner similar to a single port antenna – this can simplify the measurement procedures and can also serve to reduce the overall measurement time.

One of the problems with this type of measurement is that a power loss will be evident from any power divider that is external to any physical array. Therefore, this power loss requires quantifying and 'de-embedding' from the array to accurately deduce the array performance alone. In this chapter, a direct method of de-embedding the power divider by a new modified version of the standard single port efficiency equation is presented, which can simplify the de-embedding process.

This chapter begins with a discussion on the multi-port antennas for MIMO applications in the following subsection.

6.2   Multi-port Antennas for MIMO Applications

It is not the intention of this chapter to present specific details on the design of multi-port antennas for MIMO applications. Rather, the purpose instead is to document practical procedures in the RC of how these antennas can be characterised and the performance merits that can be typically obtained. Hence, in this chapter the antennas will only be briefly introduced.

The antenna Under Test (AUT) to be used in this work is a multi-port antenna with two feeds that is based upon a Planar Inverted F Antenna (PIFA) topology. The antenna is referred to as the 'Single Element Dual Feed PIFA'. The antenna was developed at the University of Liverpool in 2010 [6]. The motivation for the antenna development was to create a small-sized MIMO/diversity antenna based on the popular PIFA topology that could be successfully employed into a handset device.

The single element dual feed PIFAs under investigation consist of two separate cases – one where the feeds are aligned parallel to each other (co-polarised feeds); the other has the feeds aligned perpendicular to each other (cross-polarised feeds). For both antennas, the copper radiating top plate has dimensions width=40 mm and length=20 mm and ground plane dimensions of width=40 mm and length=100 mm. The dielectric material used between the rectangular ground plane and the feeds is FR4, which has a thickness of 1.5 mm and a relative permittivity ($\varepsilon_r = 4.4$). The height of the top plate from the FR4 substrate is 10 mm.

Figures 6.1 and 6.2 depict the topology of the two antennas under investigation in this work.

Shorting pin
Feed 1
Feed 2

Figure 6.1   Dual feed PIFA with co-polarised feeds.

The antennas are designed to operate in the 2.45 GHz Bluetooth/WLAN band as a diversity/MIMO antenna. The electrical separation between the two feeds evident in Figures 6.1 and 6.2 is 0.17λ at 2.45 GHz.

The isolation between the two closely spaced feeds is provided in this case by etching away a portion of the ground plane between the feeds. In this, it aims to provide a barrier to current flow by means of an antiresonant LC circuit in between the two ports to allow for an increase in isolation. The etching technique is depicted in Figure 6.3,

Shorting pin
Feed 1
Feed 2

Figure 6.2   Dual feed PIFA with cross-polarised feeds.

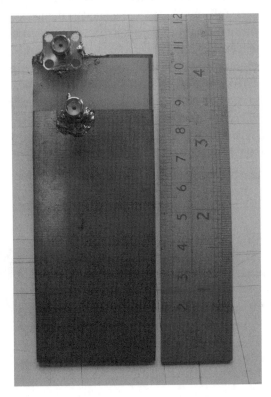

Figure 6.3   Etching technique to provide isolation between the two feeds.

while the approximate equivalent circuit and current distributions that support the technique can be viewed in Ref. [7].

There are other examples in literature to accomplish the isolation of closely spaced feeds; for example, by the insertion of a suspended line between a PIFAs feeding or shorting points [8], or via the use of de-coupling networks [9, 10].

One of the unique features and novelties of the designs and this work is that the antennas are configured only to provide pattern diversity (co-polarised feeds) and pattern and polarisation diversity (cross-polarised feeds) as opposed to spatial diversity. A brief summary of these techniques is presented as follows.

Pattern or angle diversity involves multi-port antennas that are configured to produce beams pointing in slightly different directions. This technique is able to provide diversity as it has been stated that the scattered signals associated with these (different) directions resulting from a multi-path propagation environment are uncorrelated [11].

Polarisation diversity advocates that multiple feeds on a single device to be configured with different polarisations. This technique is clever in the sense that the different (orthogonal) polarisations can serve to minimise the space requirements on a device as it capitalises on the fact that the field orthogonality suffices to de-correlate any signals [11]. Other diversity schemes do exist such as frequency and time diversity. A further in-depth discussion on all the diversity schemes can be found in Ref. [11].

With respect to the radiation pattern characteristics of the devices, Figures 6.4 and 6.5 portray the three-dimensional pattern (diversity) for the co-polarised and cross-polarised feeding networks, respectively. It can be seen that (pattern) diversity is achieved due to the fact that feed 1 mainly uses the top plate as the radiating element, whereas feed 2 uses the top plate and the ground plane as the radiating elements [5].

(a)   (b)

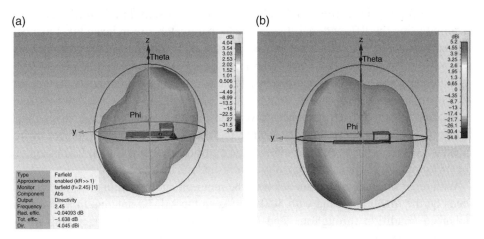

Figure 6.4 Three-dimensional simulated radiation patterns for co-polarised PIFA: (a) Feed 1 and (b) Feed 2.

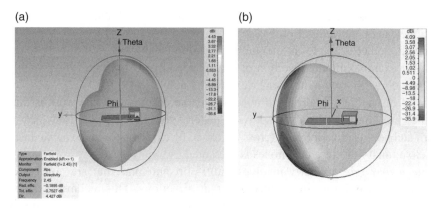

Figure 6.5 Three-dimensional simulated radiation patterns for cross-polarised PIFA: (a) Feed 1 and (b) Feed 2.

Hence, the radiation patterns produced by the two feeds are different and the diversity is aimed to be achieved through these differing patterns.

6.3 Measurement Parameters

In this work, the following performance indicators are sought:

1. Diversity gain
2. Correlation levels
3. Channel capacity
4. Embedded element efficiencies

The diversity gain, correlation levels and channel capacity in this work have been acquired using the RC, and their measurement parameters are indicated in Table 6.1. The embedded element efficiency has been acquired using an anechoic chamber and will be discussed separately in Section 6.7.

With respect to the RC measurements, Table 6.1 details the parameters used in this investigation.

In Table 6.1, the stirring sequences have been selected to allow for a relatively large amount of data to acquire the performance indicators accurately. The number of frequency data points has been selected to excite a sufficient number of modes throughout the measurement range as theoretically established in Chapter 2.

With regards to the miscellaneous row, the requirement for terminating any unused feeds on the MIMO antenna with 50 Ω is to keep the MIMO antenna in a consistent balanced condition throughout, and to prevent any differences occurring in the mismatch from either feed if they were left open circuit [12].

This aspect should always be followed during a measured sequence for multi-port antennas.

Table 6.1   Measurement parameters for MIMO investigation.

Parameter	Description
Frequency (MHz)	2000–3000
Stirring sequences	1 degree mechanical stirring
	Polarisation Stirring
	12.5 MHz frequency stirring
Number of frequency data points	801
Source power (dBm)	−7
Reference antenna	Satimo SH2000
Transmitting antenna	Rohde & Schwarz HF 906
Miscellaneous	MIMO antenna port not being measured, terminated in 50 Ω
	Any other antenna not being measured (e.g. reference) left inside the chamber and terminated in 50 Ω.

The requirement for having all antennas inside the chamber at the same time has been previously established and is to ensure that the chamber Q factor is constant from the reference measurement to the MIMO antenna measurement to ensure accuracy.

This aspect should always be followed during a measured sequence for all antenna measurements.

6.4   Diversity Gain from Cumulative Distribution Functions (CDF)

Before any results are issued, a few definitions are warranted.

In the simplest terms, the diversity gain can be expressed as the improvement in the Signal to Noise Ratio (SNR) at the output of a diversity combiner compared to the input of the diversity antenna at a given (usually 1%) cumulative probability level [13]. A more precise definition is possibly based on the type of diversity gain one is interested in Ref. [14]:

1. **Apparent diversity gain:** Difference between the power levels in decibel scale at a given cumulative probability level between the CDF of a combined signal and the CDF of a signal at the antenna port with the strongest average signal level.
2. **Effective diversity gain:** Difference between the power levels in decibel scale at a given cumulative probability level between the CDF of a combined signal and the CDF of a signal at the port of an ideal single antenna corresponding to 100% radiation efficiency when measured in the same environment.

3. **Actual diversity gain:** Difference between the power levels in decibel scale at a given cumulative probability level between the CDF of a combined signal and the CDF of a signal at the port of an existing practical single antenna that is to be replaced by the diversity antenna when measured at the same location and in the same environment.

In this chapter we will deduce the results with reference to the apparent diversity gain scheme. With regards to the diversity combination techniques, many different schemes have been devised to exploit the (ideal) uncorrelated fading exhibited by separate isolated antenna elements.

In this work, the Selection Combining (SC) technique is to be applied with respect to all diversity gain results that are issued; the reason for this selection is simply because from a signal processing and hardware perspective, it is perhaps one of the easiest methods of all to apply [14].

The basic premise of the SC technique will be outlined in this work, followed by a brief statement of other existing techniques. For further in-depth analysis of the fundamental principles of all existing combination techniques, please refer to Ref. [11].

Figure 6.6 illustrates the principle of selection combining.

As Figure 6.6 aims to illustrate the signal with the highest SNR is simply selected. This is one example of how multi-path fading can be overcome – that is, if a signal received by one antenna suffers from fading, any one of the other antennas should receive a signal with a larger power (provided they are sufficiently uncorrelated), and the largest signal among all receivers is thus selected.

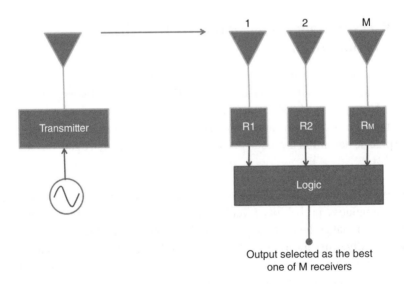

Figure 6.6   Principle of selection combining.

Other combinational techniques include Maximal Ratio Combining (MRC) and Equal Gain Combining (EGC).

With reference to Jakes [11], *Maximal Ratio Combining* requires that the M signals are weighted proportionally to their signal voltage to noise power ratios and then summed. In contrast to the SC technique, the M signals must be co-phased before combination takes place [11].

However, it may not always be convenient or desirable to perform the variable weighting of the signals that is required for MRC. If this is the case, *Equal Gain Combining* can then be performed which advocates that all the 'gains' are set to a constant value of unity and the combined output is taken as a summation of all branches (inputs to the combiner) [11].

Using the RC, it is possible to calculate the diversity gain from both CDFs and the measured correlation between two or more antenna feeds. In this subsection, the CDF approach will receive treatment first.

To begin with, a transfer function from a reference antenna measurement is established as detailed in Equation 6.1. The purpose of this measurement is to 'calibrate' the attenuation in the chamber. This should be performed similar to the earlier defined procedures.

$$\mathrm{TF_{REF}} = \frac{\left\langle \left|S_{21\mathrm{REF}}\right|^2 \right\rangle}{\left(1 - \left(\left|S_{11\mathrm{TX}}\right|\right)^2\right)\left(1 - \left(\left|S_{22\mathrm{REF}}\right|\right)^2\right)} \tag{6.1}$$

where $|S_{11\mathrm{TX}}|$ = reflection coefficient of the transmitting antenna and $|S_{22\mathrm{REF}}|$ = reflection coefficient of the reference antenna. Please note that similar to previous chapters, the reflection coefficient quantities are acquired in an AC and are not signified as being from an ensemble average. The reference transfer function in (6.1) should then be corrected for its known radiation efficiency (η_{REF}) as in (6.2).

$$P_{\mathrm{REF}} = \frac{\mathrm{TF_{REF}}}{\eta_{\mathrm{REF}}} \tag{6.2}$$

Once completed, the reference antenna should then be substituted for the MIMO antenna.

The transmission coefficient is then measured between the transmitting antenna and both feeds of the antenna (in turn), ensuring that the feed not undergoing measurement is terminated in a matched load (50 Ω in this case). The resulting channel samples ($S_{1\times2}$); that is, one transmitting and two separate receive channels (Single Input Multiple Output; SIMO) in our case, are a function of frequency and the stirring parameters as detailed in Table 6.1. If multiple transmitting antennas exist in the chamber, then this measurement process should be repeated with respect to each channel in turn.

It is necessary to then normalise the measured channel samples with respect to the square root of the pre-defined reference power level. The square root is needed because the measured channel samples ($S_{1\times 2}$) are a function of voltage (measured transmission coefficient S-parameters), and the normalisation takes into account the path loss in the chamber, defined by the reference transfer function and the known radiation efficiency in (6.2). Thus, the resulting channel matrix ($H_{1\times 2}$) can be formed by [15].

$$H_{1\times 2} = \frac{S_{1\times 2}}{\sqrt{P_{REF}}} \tag{6.3}$$

Once the channel matrix is formed, the cumulative probability can be calculated by Equations 6.4 and 6.5, and is graphed against the relative received power, calculated by Equations 6.6–6.8.

For the cumulative probabilities, calculate the sum for each respective channel (i):

$$\text{sum}\_i = \sum_{n=1}^{N} H_{1\times i} \tag{6.4}$$

where $i = 1, 2$ and $N = 718$. After which, the cumulative probability (Cumprob) can be deduced from:

$$\text{Cumprob}\_i = \frac{\text{Cumsum}\_i}{\text{sum}\_i} \tag{6.5}$$

where Cumsum represents the cumulative sum.

For the relative received power, the channel matrix (voltage) value for each respective branch is squared:

$$H_{sq}\_i = \left(H_{1\times i}\right)^2 \tag{6.6}$$

The time average is then found from (6.7).

$$T_{AV}\_i = \frac{1}{N}\sum_{n=1}^{N} H_{sq}\_i \tag{6.7}$$

Each power sample is then normalised to the time average and converted to decibel scale by (6.8).

$$P_i(\text{dB}) = 10\log_{10}\left(\frac{H_{sq}\_i}{T_{AV}\_i}\right) \tag{6.8}$$

The time average is required in this case because the measured samples are subject to fading in the channel (i.e. RC measured magnitudes are acquired in a non-line of sight (NLoS) Rayleigh environment). This necessitates that the samples be averaged over the statistical distribution in the channel.

To calculate the combined level in the CDF plots, a decision is made with respect to the normalised channel samples in (6.3). For example, for the SC combination technique, the magnitudes of the channel samples are compared at this stage and the maximum is selected as previously stated which subsequently forms a new array of values. This separate array is then processed alongside the measured channel samples (for the MIMO antenna ports) following the steps defined in Equations 6.4–6.8.

For different diversity combination schemes, it is a decision of how to treat the measured channel samples from (6.3), which will determine the combined CDF level from which to compare the magnitude of apparent diversity gain.

Figures 6.7 and 6.8 represent the measured apparent diversity gain from CDFs for the co-polarised PIFA and cross-polarised PIFA, respectively.

From Figure 6.7 it can be seen that a high level of apparent diversity gain is evident from the co-polarised design. The power level of the strongest branch at the 1% cumulative probability level (Channel 2) is recorded as −22.104 dB, with the power level of the selection combined level being measured at −12.66 dB. This provides an apparent diversity gain of 9.44 dB.

From Figure 6.8 it can be seen that a high level of apparent diversity gain is obtained for the cross-polarised design. The power level of the strongest branch at the 1% cumulative probability level (Channel 1) is recorded as −21.086 dB, with the power

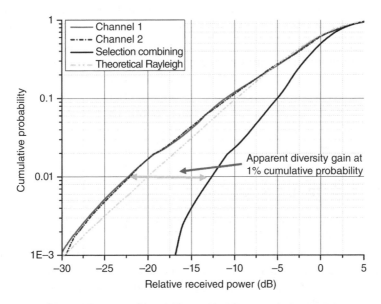

Figure 6.7 Diversity gain from CDF for co-polarised PIFA.

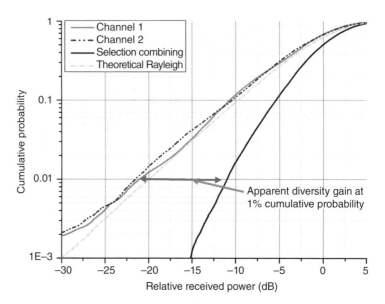

Figure 6.8   Diversity gain from CDF for cross-polarised PIFA.

level of the selection combined level being measured at −10.991 dB. This provides a
diversity gain of 10.095 dB; very close to the theoretical two-port selection combining
a maximum of 10.2 dB in this case.

Both channel branches for co-polarised and cross-polarised designs are seen to follow the theoretical Rayleigh CDF curve well which would suggest that the samples are
inherently Rayleigh distributed. However, they are slightly offset to the left of the
Rayleigh curves, which is due to the fact that the respective branch efficiencies are less
than 100% [2]. The theoretical Rayleigh curve here signifies the performance of an
ideal lossless (and hypothetical) antenna with 100% efficiency.

So far the evidence would suggest that the antennas perform well, and the isolation
technique employed would appear successful in eliminating any detrimental performance effects. However, we seek further evidence to verify the antennas' operational
quality and to substantiate the measured levels of diversity gain witnessed.

6.5   Diversity from Correlation

Although the diversity gain can be easily predicted from the measured and calculated
CDFs, a degree of information is lost here as all the measured samples for each channel
across the entire frequency range have been utilised in order to define the CDFs
accurately. In many instances, it could be important to know how the diversity gain
varies as a function of frequency. To achieve this, the correlation between measured
samples from both the antenna feeds can be measured and calculated as a function of
frequency, and the diversity gain thus obtained.

It is well known that the complex correlation coefficient (ρ) can be measured and calculated from the far field embedded element pattern acquired in an AC [14]:

$$\rho = \frac{\displaystyle\iint_{4\pi} G_1(\theta,\phi) \cdot G_2^*(\theta,\phi) d\Omega}{\displaystyle\iint_{4\pi} G_1(\theta,\phi) \cdot G_1^*(\theta,\phi) d\Omega \iint_{4\pi} G_2(\theta,\phi) \cdot G_2^*(\theta,\phi) d\Omega} \tag{6.9}$$

where $G_1(\theta,\varphi)$ and $G_2(\theta,\varphi)$ are the embedded far field functions of ports 1 and 2, respectively – that is, the far field function when all the other elements are present and terminated in impedance matched loads, and (*) represents the complex conjugate. It is also possible to express the same complex correlation in terms of S-parameters measured at the antenna ports [14].

$$\rho = \frac{S_{11}^* S_{12} + S_{21}^* S_{22}}{\left[1 - \left(|S_{11}|^2 + |S_{12}|^2\right)\right] \times \left[1 - \left(|S_{21}|^2 + |S_{22}|^2\right)\right]} \tag{6.10}$$

where the square of the complex correlation in (6.10) is equal to the envelope correlation. The form of (6.10) is only valid for lossless antennas; hence it will not be employed here. Using the same RC measured data as in Section 6.4, the correlation between both antenna feeds can be found from the measured voltage samples ($S_{1\times2}$) by use of (6.11) after consultation with [16].

$$\rho = \frac{S_{xy}}{S_x S_y} \tag{6.11}$$

where x represents the samples obtained for feed 1 in this case and y represents the samples obtained for feed 2. Also:

$$S_{xy} = \frac{1}{N-1} \sum_{j=1}^{N} (x_j - \bar{x})(y_j - \bar{y}) \tag{6.12}$$

where $\bar{x} = \frac{1}{N} \sum_{j=1}^{N} x_j$ and $\bar{y} = \frac{1}{N} \sum_{j=1}^{N} y_j$

$$S_x^2 = \frac{1}{N-1} \sum_{j=1}^{N} (x_j - \bar{x})^2 \tag{6.13}$$

and

$$S_y^2 = \frac{1}{N-1} \sum_{j=1}^{N} (y_j - \bar{y})^2 \tag{6.14}$$

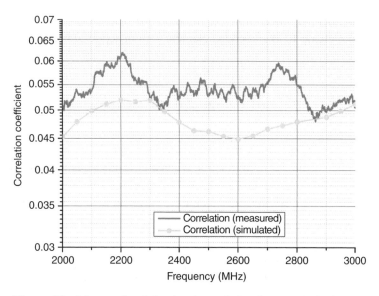

Figure 6.9   Measured and simulated correlation for co-polarised PIFA.

Figures 6.9 and 6.10 detail the measured and simulated correlation as a function of frequency for the co-polarised and cross-polarised PIFA, respectively. The simulated correlation values have been acquired in CST Microwave Studio from the optimised antenna models developed in Ref. [6]. The correlation (and the subsequent diversity gain) values have been calculated from the three-dimensional far field embedded patterns (Eq. 6.9) and not from simulated S-parameters (Eq. 6.10) for reasons previously explained.

From Figures 6.9 and 6.10 it can be seen that the correlation levels, measured from voltage samples obtained at both antenna ports, are very low indeed. Both figures prove that the respective feeds for both the co-polarised and cross-polarised designs are well isolated from one another, and as such, the respective voltage samples obtained at either port are substantially different. The agreement with simulated values is also seen to be very good; generally, differences in the order of 0.01 are witnessed.

The low values of measured correlation in Figures 6.9 and 6.10 potentially validate the diversity gain results obtained from the CDFs, as a low correlation between the antenna ports can give rise to a high level of diversity gain as we can see next.

Mathematically, the standard definition for diversity gain (DG), estimated from the complex correlation goes as [1].

$$DG = 10\sqrt{1 - |\rho|^2} \qquad (6.15)$$

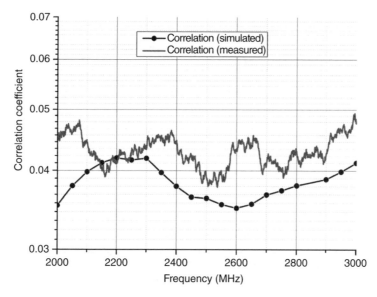

Figure 6.10   Measured and simulated correlation for cross-polarised PIFA.

where $\sqrt{1-|\rho|^2}$ is the approximate expression for the correlation efficiency – that is, the reduction in diversity gain due to the correlation between the two antenna feeds. Assuming (two port) selection combining techniques, it is possible to revise (Eq. 6.15) to [17]:

$$DG = 10.48\sqrt{1-|\rho|^2} \tag{6.16}$$

where the 10.48 in (Eq. 6.16) is quickly proved as follows:

$$\left.\begin{array}{l} 0.01 = 1 - e^{-SNR} \Rightarrow SNR = -\ln\,(0.99) = 0.01 = -20 \text{ dB} \\[2mm] 0.01 = 1 - \left(e^{-SNR}\right)^2 \Rightarrow SNR = -\ln\left(1 - \sqrt{0.01}\right) = 0.105 = -9.8 \text{ dB} \end{array}\right\} 10.2 \text{ dB} = 10.48$$

$$\tag{6.17}$$

Figures 6.11 and 6.12 detail the measured and simulated diversity gains which is calculated from the correlation using (6.16), for the co-polarised and cross-polarised designs, respectively.

From Figures 6.11 and 6.12 we see that very high levels of diversity gain are witnessed due to the very low correlation levels between both feeds on each respective antenna. The (average) difference in diversity gain between the CDF deduced and the correlation deduced results for the co-polarised design is 0.76 dB, while for the

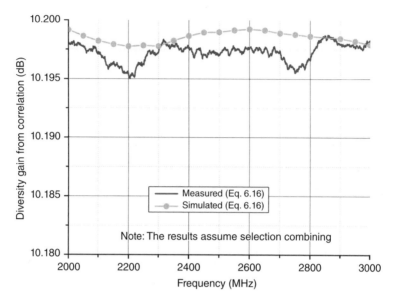

Figure 6.11   Measured and simulated diversity gain from correlation for co-polarised PIFA.

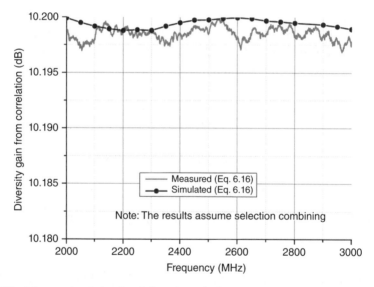

Figure 6.12   Measured and simulated diversity gain from correlation for cross-polarised PIFA.

cross-polarised design the difference is 0.103 dB. This proves that from either method (CDF or correlation deduced), the results are generally comparable to one another. Further, the results from the correlation analysis substantiate the accuracy in the CDF measurement results, proving its appropriateness and correctness. To conclude this subsection, it is proven therefore that small antennas with closely spaced feeds that cannot employ spatial diversity can perform successfully in a MIMO operational role.

6.6 Channel Capacity

The channel capacity is a measure of how many bits per second can be transmitted through a radio channel per Hertz – a quantity which can also be referred to as the spectral efficiency [1]. By the use of the same normalised channel samples from the diversity gain measurement (Eq. 6.3), and assuming that the receiver has perfect channel state information, the channel capacity can be deduced from the measured channel matrices by [15]:

$$C = \left\langle \log_2 \left\{ \det \left(I_{1\times 2} + \frac{\text{SNR}}{N_{\text{TX}}} \left(H_{1\times 2} \right) \left(H_{1\times 2}^T \right) \right) \right\} \right\rangle \tag{6.18}$$

where I = identity matrix, N_{TX} = number of transmitting antennas = 1 in this case, and (^{T}) signifies the complex conjugate transpose. The average notation in (6.18) is required because the channel capacity and SNR are subject to fading with time in the RC. Hence, the capacity needs to be averaged over the statistical distribution in the channel [14].

Figure 6.13 details the measured 1×2 (SIMO) channel capacity for the co-polarised and cross-polarised designs. In Figure 6.13, the measured result is compared against a

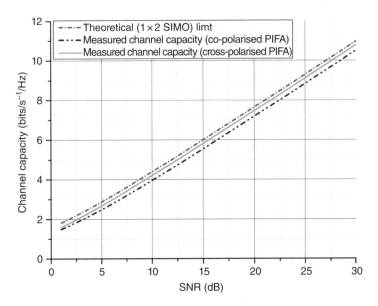

Figure 6.13 Measured and theoretical channel capacity as a function of SNR.

maximum theoretical capacity that assumes that the two receive ports are completely uncorrelated. The maximum theoretical capacity for a parallel 1×2 (SIMO) channel can be calculated from Ref. [14]:

$$C_{1\times 2} = \log_2\left(1 + SNR_{TX1\_RX1} + SNR_{TX1\_RX2}\right) \tag{6.19}$$

From Figure 6.13 it can be seen that the two designs yield a high level of capacity owing to the low correlation levels. The cross-polarised design is seen to exhibit slightly higher levels of capacity – this can be attributed to the slightly reduced correlation levels between the two antenna ports as proven in Figures 6.9 and 6.10.

6.6.1 General Problems to Avoid: Statistical Variations

In Ref. [15], a discussion is presented concerning the accuracy of the channel capacity model in (6.18) – referred to as the 'full correlation model' against an alternative channel format referred to as the 'Kronecker model'. The basic difference between the two formats is in the construction of the channel matrix – the Kronecker model relies on the use of a covariance matrix format which differs slightly from the adopted procedures used throughout this chapter.

In Ref. [15] it is shown that the two different models can predict the same capacity level; however, when using a small-sized RC, the capacity could be overestimated using both models at high SNRs and for more than three receive elements. It was concluded that the subsequent reason for the overestimation was that the measured channel samples did not satisfy Multivariate Normality (MVN) – that is, the measured channel samples were not jointly normally distributed, contravening the single antenna normal statistical model as discussed in Chapter 2.

In Appendix A, the normality of the two receive channels is put to the test to provide confidence in the estimation of the channel capacity for both designs in Figure 6.13. This aspect represents a subtle issue to be aware of and something that should be tested to ensure that no statistically significant variations exist. Appendix A also demonstrates a means of going about this testing process, which is referenced back to the statistical theory presented in Chapter 2.

6.7 Embedded Element Efficiency

The embedded element efficiency – that is, the efficiency of the MIMO antenna when one port is excited and the unused ports are terminated in impedance matched loads, can be expressed and treated simply by the use of S-parameters.

The quantities derived here are based on the total embedded radiation efficiency – that is, mismatch factors due to imperfect impedance matching are included. It is stated in Ref. [14] that if the diversity antenna is constructed from materials with small ohmic losses, then the major contributions towards the total radiation efficiency are from the reflections on the excited port and the absorption of power in the impedance matched terminations.

The difference between the power that is accepted by the excited port (P_{ACC}) and the absorption in the impedance matched ports equates to the total radiated power (P_{RAD}). From Ref. [14], which assumed the use of a two-port network model, the power dissipated on the impedance matched port (assumed as port 2) can be expressed as:

$$\eta_{ABS} = \frac{P_{RAD}}{P_{ACC}} = \frac{1 - |S_{11}|^2 - |S_{12}|^2}{1 - |S_{11}|^2} \tag{6.20}$$

The power accepted by the excitation port (assumed as port 1) can be stated as [14]:

$$\eta_{MISMATCH} = 1 - |S_{11}|^2 \tag{6.21}$$

Thus the total embedded element efficiency is expressed as [14]:

$$\eta_{TOT\_EMBED} = \eta_{ABS} \times \eta_{MISMATCH} = 1 - |S_{11}|^2 - |S_{12}|^2 \tag{6.22}$$

From (6.22), using the assumption that the ohmic losses in the construction materials are small and excitation port is well impedance matched, the dominant contribution towards the efficiency is from the mutual coupling between the antenna ports. That is, the higher the mutual coupling, the more power will be dissipated in the loads of the other antenna ports and lower the resultant efficiency will be. It is also stated from Ref. [14] that the mutual coupling term represents a fundamental limitation in conventional antenna arrays – that is, if the inter-element spacing is small (typically <0.5 wavelengths), the mutual coupling can offer a severe reduction in efficiency [18].

An additional efficiency factor should be present in (6.22) if the ohmic losses are to be considered. In the results to follow, the assumption that the ohmic losses are small has been applied, which is considered reasonable since the antennas are constructed from high-quality copper having a large material conductivity ($\approx 5.7 \times 10^7$ S/m at 2.45 GHz).

Figures 6.14 and 6.15 detail the measured and simulated S-parameters for the co-polarised and cross-polarised design, respectively, while Figures 6.16 and 6.17 detail the measured and simulated embedded element efficiencies. All measured parameters herewith were acquired in an anechoic and all unused ports in the measurements have been terminated in impedance matched loads (50 Ω in this case).

From the measured S-parameters (Figure 6.14b) for the co-polarised design it can be seen that the antenna is reasonably well isolated ($S_{21} \approx -10$ dB) at the desired centre frequency. As we have already seen, this is sufficient for the measured correlation

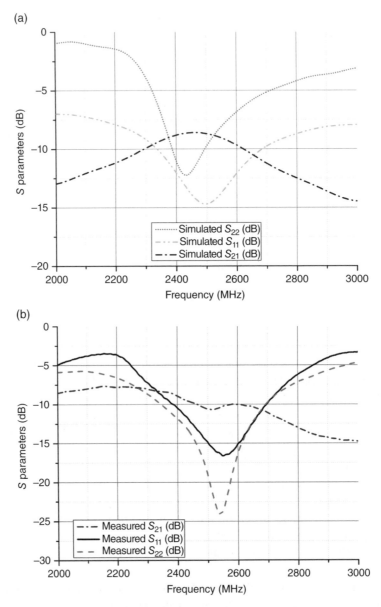

Figure 6.14 (a) Simulated S-parameters for co-polarised PIFA and (b) measured S-parameters for co-polarised PIFA.

between both feed ports to be acceptably low. Both antenna ports are also seen to be well impedance matched at the desired centre frequency, meaning that the proportion of power lost due to impedance mismatch about the centre frequency should be relatively small.

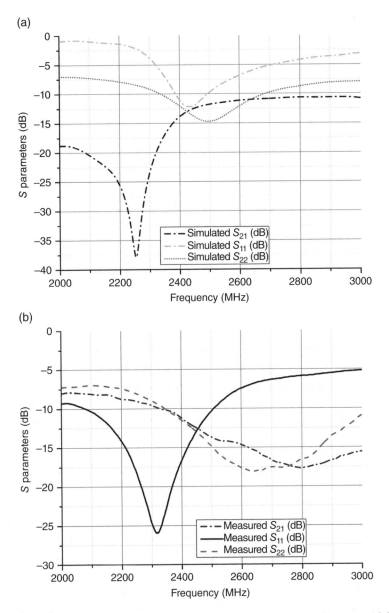

Figure 6.15 (a) Simulated S-parameters for cross-polarised PIFA and (b) measured S-parameters for cross-polarised PIFA.

From the measured S-parameters (Figure 6.15b) for the cross-polarised design it can be seen that the antenna is slightly better isolated ($S_{21} \approx -12.5$ dB) at the desired centre frequency than the co-polarised design (Figure 6.14b). Measured data for port 1 is seen to be slightly offset from the centre frequency which can be attributed to slight

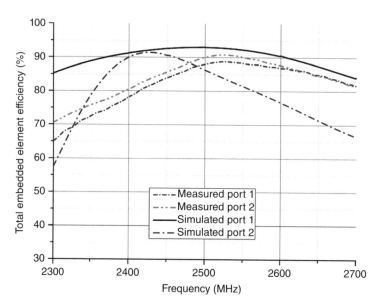

Figure 6.16   Measured and simulated total embedded element efficiency for co-polarised PIFA.

differences in dimensions between the simulated and manufactured design – this is because the practical cross- polarised design was manufactured in three separate pieces and the placement of the orthogonal feed was not exact enough. Nevertheless, at the centre frequency it is still well impedance matched. On the whole, both designs are more than acceptable for operation.

The measured and simulated total embedded element efficiency (including matching losses) for the co-polarised design can be viewed in Figure 6.16. Please note that the graph has been constrained about the centre frequency to more clearly depict the measurement and simulation comparison at the antennas' desired working range.

From Figure 6.16 it can be seen that the co-polarised antenna yields a high level of efficiency – the measured levels at the peak are at 91%. The agreement with the simulated assessment is very good at the peak values – away from the peak value; the simulated assessment for port 1 is much higher than the measured data over the first 200 MHz. This is due to the greater impedance matching in this range from the simulated design. Conversely, the simulated assessment for port 2 is a lot lower than the measured data owing to the worse impedance match in this range.

From Figure 6.17 it can be seen that the cross-polarised antenna also yields a high level of efficiency – the measured levels at the peak for branch 1 is 90%, and 96% for branch 2. The agreement with the simulated assessment is very good around the centre frequency – away from this the simulated assessment for port 1 is much lower than the measured data over the first 100 MHz. This is due to the greater impedance matching

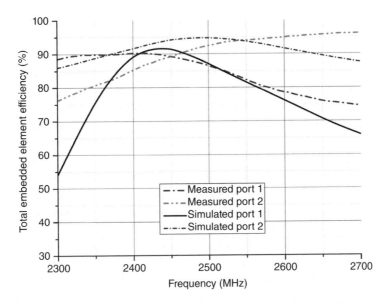

Figure 6.17 Measured and simulated total embedded element efficiency for cross-polarised PIFA.

in this range from the measured design. Conversely, the simulated assessment for port 2 is a lot higher than the measured data owing to the better impedance match in this range; on the whole though the agreement is reasonable. To improve this agreement it necessitates that the tolerances in the manufacturing strategy always needs to be as precise as possible.

6.8 Definitions: Conventional Array Antenna Measurements

As previously stated, for multi-port arrays that are not designed for MIMO applications, a distinction should perhaps be created, as in their practical operation they are typically used in an 'all excited' manner – that is, we are not necessarily interested in separate 'embedded' channels. The rest of this chapter is devoted to the efficiency characterisation of larger sized (conventional) antenna arrays designed for radio astronomy applications.

In this chapter, a new equation is developed to characterise the efficiency performance of conventional arrays, taking into account the realistic 'all excited' nature of the arrays' practical operation. This subsection is concerned with an introduction of the antenna array under test. This work results from a collaboration between two parties; the antenna array (denoted by the Octagonal Ring Antenna or ORA for short), was developed by A. K. Brown and Y. Zhang at the University of Manchester [19].

The antenna under test in this work is a five-element compact dual polarised aperture array prototype based on an octagonal ring configuration as proposed in Ref. [19]. The basic design premise of the array has been formed from the conceptual theory of current sheet arrays (CSA) introduced by Wheeler [20]. Thus, the design aims to make use of, instead of reducing the mutual coupling between the array elements. The reason why the coupling is required in the design strategy is because in order to keep the radiation pattern side lobes under control, the array must be operated in a region where the electrical separation between elements is small; hence the mutual coupling is high. This also necessitates why the array must be used in an 'all-excited' manner.

The spacing between successive elements in the prototype is configured in a triangulated manner with the horizontal distances set at 112 mm between each consecutive feed and vertical feedlines of 55 mm in length. Each antenna element includes an integrated stripline transition which was known to introduce some loss. However due to its integrated nature this was treated as part of the antenna. The desired operational frequencies of the prototype are from 400 to 1400 MHz in which the aforementioned coupling effects are used to obtain a good level of impedance matching throughout the entire range. The overall dimensions of the array are width = 540 mm, height = 215 mm and the depth of the array from the front to the metallic ground at the back is 105 mm.

The front and side view of the array prototype can be viewed in Figure 6.18a and b, respectively.

6.9   Measurement Parameters

The measurement parameters and procedures will be split into two sections; one to detail the all-excited efficiency measurements in the RC, the other details the power loss deduction of the power dividers that have been employed in this work.

6.9.1   RC 'All-Excited' Measurement Parameters

The stirring sequences are configured to encompass one mechanical stirring intervals and polarisation stirring. A total of 718 measurement samples exist per frequency point. The measured frequency ranges in this instance were selected from 400 to 1000 MHz owing to the upper frequency limit of the power divider employed. A total of 801 frequency data points was utilised to ensure a sufficiently large number of modes would be excited in the chamber throughout the measurement range. As consistent with standard efficiency measurement procedures in the RC, a reference measurement first took place for calibration purposes; a log-periodic antenna (Rohde & Schwarz HL223) was used having known performance values. The transmitting antenna was a homemade Vivaldi antenna working from 400 MHz to 2 GHz. The set-up used in the all-excited array measurement is depicted in Figure 6.19a and b.

(a)

(b)

Figure 6.18   (a) Front view of five element ORA prototype and (b) side view of five element ORA prototype.

6.9.2   *Power Divider Measurement Procedures*

For the determination of the power divider loss values, an open air test site was selected. An 8:1 power divider (Mini Circuits model no. ZC8PD1-10-S+) with a Voltage Standing Wave Ratio (VSWR) < 1.21 from 300 MHz to 1 GHz was employed in this

(a)   (b)

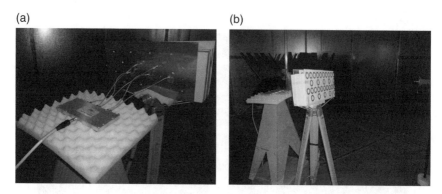

Figure 6.19   Measurement set-up: (a) power divider connections and (b) array mounting.

study. The power divider had rigid coaxial cables connected from the output ports, of which five in total were used to connect to the array feeds with equal weights. This left three ports on the power divider obviously unused; hence throughout all measured sequences these ports were terminated in impedance matched loads (50 Ω in this case). Figure 6.20 depicts the measurement set-up to determine the power divider loss.

To deduce the loss of the power divider accurately, the same impedance matched loads on the unused ports need to remain in place during the loss deduction measurement to ensure the power dissipated by the divider is consistent.

To determine the power loss, S-parameters can be employed in conjunction with a vector network analyser (VNA). All other outputs except the testing port are terminated in impedance matched loads and the transmission coefficient $|S_{21}|$ is measured between the common input and one output as a function of frequency. This gives rise to five separate transmission coefficient measurements; one for each output port, and a total insertion loss (T_{IL}) is defined according to (6.23).

$$T_{IL}\left(dB\right)=10\log_{10}\left\{\sum_{m=2}^{6}\left|S_{m,1}\right|^{2}\right\}$$ (6.23)

where $\left|S_{m,1}\right|=\left(V_{R,m}/V_{T}\right)$, for $m=2,\,3,\,4,\,5,\,6$, respectively.

6.10   Deduction of Characterisation Equation

As previously stated, the presence of the power divider will present a power loss that is external to the physical array under test. In this section, the direct de-embedding technique is developed that is used in this study to account for the loss and calculate the array's efficiency in one.

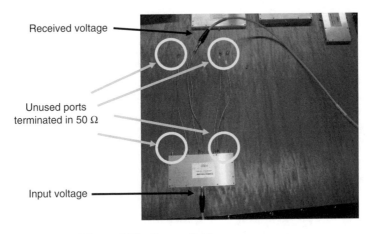

Received voltage

Unused ports
terminated in 50 Ω

Input voltage

Figure 6.20   Power divider measurement.

In standard single port efficiency measurements (from Eq. 5.7), assuming that the transmitting antenna (excited from port 1 in this case) is well impedance matched, the power transfer function for the reference and antenna under test at the receiving side can be determined by the well-known power ratio (P) in (6.24).

$$P = \frac{\left\langle |S_{21}|^2 \right\rangle}{\left(1 - \left(|S_{22}|\right)^2\right)}$$

(6.24)

where the reflection coefficient parameter in (6.24) is assumed to be measured in an AC, $\langle\ \rangle$ = average of the scattering parameters and the term $\left(1 - \left(|S_{22}|\right)^2\right)$ is applied to effectively normalise the averaged power transmission coefficient by a factor of what power is reflected from the terminal due to impedance mismatch.

Clearly, the inclusion of the power divider will offer a further power loss in addition to (6.24) before the cable on the receiving side (Figure 6.19a). Therefore, a further normalisation factor is simply introduced to account for the loss which has nothing to do with the array. Hence, when the array is introduced, the normalisation factor in (6.24) effectively becomes (6.25).

$$\left(1 - \left(|S_{22}|\right)^2\right) \times \left(PD_{Loss}\right)$$

(6.25)

where (PD_{Loss}) is the linear form of (T_{IL}) from (6.23). The complete unknown radiation efficiency for the antenna array (η_{RAD}) with the power divider then becomes as (6.26)–(6.29) using the same adopted terminology.

$$P_{REF} = \frac{\left\langle |S_{21REF}|^2 \right\rangle}{\left(1 - \left(|S_{22REF}|\right)^2\right)} \tag{6.26}$$

$$P_{ARRAY} = \frac{\left\langle |S_{21ARRAY}|^2 \right\rangle}{\left(1 - \left(|S_{22ARRAY}|\right)^2\right) \times \left(PD_{Loss}\right)} \tag{6.27}$$

$$\eta_{AUT} = \left\{\frac{P_{ARRAY}}{P_{REF}}\right\} \times \eta_{REF} \tag{6.28}$$

$$\eta_{AUT} = \left\{\frac{\left\langle |S_{21ARRAY}|^2 \right\rangle}{\left\langle |S_{21REF}|^2 \right\rangle} \times \frac{\left(1 - \left(|S_{22REF}|\right)^2\right)}{\left(1 - \left(|S_{22ARRAY}|\right)^2\right) \times \left(PD_{Loss}\right)}\right\} \times \eta_{REF} \tag{6.29}$$

The total radiation efficiency of the array can be calculated from (6.30).

$$\eta_{TOTAL} = \eta_{AUT} \times \left(1 - \left(|S_{22ARRAY}|\right)^2\right) \tag{6.30}$$

where $_{ARRAY}$ and $_{REF}$ stand for the array and reference antennas, respectively and η_{REF} signifies the known reference efficiency. Consistent with the prior investigations in this book, the reflection coefficient parameters have been acquired in an AC hence the omission of the ensemble average. Further, the reflection coefficient parameter for the antenna array should be acquired in conjunction with the power divider – the mutual coupling levels on the antenna will differ from an embedded scenario to the all-excited scenario, because the power loss between successive embedded elements will not be the same throughout each excitation, owing to the differing distances between the embedded element and the single excitation location.

In this case, the difference in mutual coupling levels meant that in an embedded scenario, the mutual coupling levels were much reduced and the array would not be properly impedance matched over the entire frequency band (this was confirmed practically). Hence, to obtain the desired impedance matching, the coupling levels had to be strong which the all-excited scenario provided.

6.11   Measurement Results

In this subsection the results will be split into two separate sections for the power divider loss and the radiation efficiency performance of the array.

6.11.1   Power Divider Measurement Results

With respect to Section 6.9.2, Figures 6.21 and 6.22 detail the individually measured transmission coefficients for the five feeds of the power divider, and the combined total insertion loss (Eq. 6.23) in both linear and decibel scales, respectively.

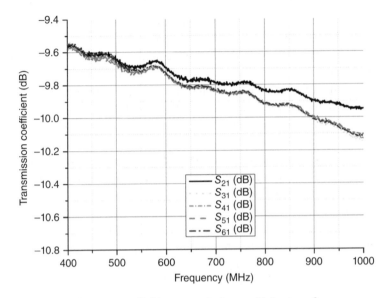

Figure 6.21   Power divider transmission coefficients vs frequency.

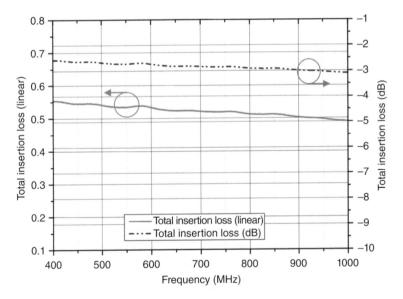

Figure 6.22   Power divider total insertion loss vs frequency.

From Figure 6.22, the linear values depicted are the values used in (Eq. 6.29) to 'de-embed' the power divider and thus deduce the array's efficiency.

6.11.2 Antenna Array Measurements Results

We can now proceed to disclose the antenna array's reflection coefficient and efficiency performance (Figures 6.23 and 6.24, respectively).

From Figure 6.23 it can be seen that when the antenna is operational in conjunction with the power (i.e. an all excited mode of operation) an excellent impedance match is obtained across the working frequency band. The designers' concept of making use of mutual coupling in this regard is seen to be fully validated.

In Figure 6.24 the efficiency quantities have been plotted with and without the 'de-embedding' performed to re-enforce the necessity of this operation.

From Figure 6.24 it can be seen that the effect of not de-embedding the power divider is significant. The de-embedding technique developed in this work is direct and straightforward to implement, and the passive 'all-excited' measurement can characterise the efficiency of the complete array in a timely manner.

The trend of the efficiency results, particularly over the first 300 MHz band, suggests that the mutual coupling is having a diminishing effect on the magnitude of the efficiency; which, since (electrically) the inter-element spacing is increasing for a decreasing wavelength, is in line completely with expectations.

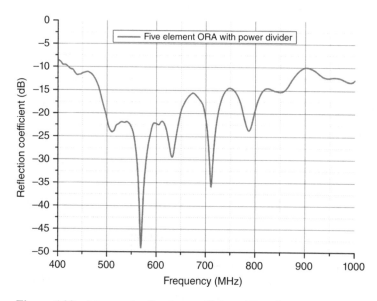

Figure 6.23 Measured reflection coefficient (dB) with power divider.

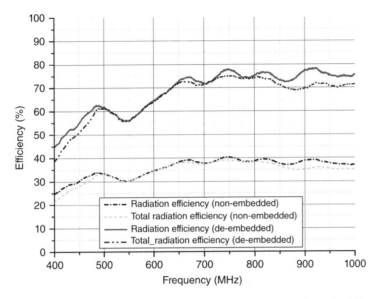

Figure 6.24   Measured array efficiencies with and without de-embedding.

No simulated assessment is provided against the measurement results in Figure 6.24 with respect to the following two points.

1. The mutual coupling in the simulated assessment, unless it can be specifically configured, would be based upon an 'embedded element' approach; which, as previously discussed, would underestimate the strength of the mutual coupling in the antenna array as opposed to the all-excited scenario. The net effect of this would be twofold: (i) the impedance matching characteristics will differ (from all-excited to embedded scenario) and (ii) because the mutual coupling levels can be lower, a larger radiation efficiency value can be predicted than what necessarily exists in practice.
2. The simulated assessment should also be performed with a large structure in order to obtain a reasonable approximation to the coupling levels in the array. Computationally speaking, this can be challenging. It has been stated that in this case, an existing simulated smaller section of the array is not big enough to obtain an accurate representation of the overall coupling, and any subsequent efficiency results could be questionable [21].

We can proceed therefore to directly assess the uncertainty in the array measurements to establish confidence in the derived results.

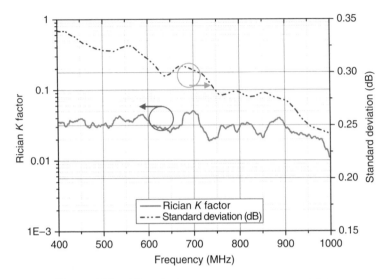

Figure 6.25   All excited array measurement uncertainty.

6.12   Measurement Uncertainty

The uncertainty models used for RC measurements have already been defined; the Rician K factor in Section 2.8 and the standard deviation in Equations 5.9 and 5.10. Therefore, to avoid repetition they will not be discussed again here. Figure 6.25 details the Rician K factor and standard deviations inherent in the array measurements.

From Figure 6.25 it can be seen that the uncertainty inherent in the measured approach is low. The Rician K factor proves that no direct proportion of power is evident in the measurement scenario and the standard deviations justify the measurement procedures and parameters selected.

6.13   Summary

This chapter has charted a distinction between multi-port antennas designed for MIMO applications and conventional antenna arrays that are not. For the MIMO applications, two new and novel single element dual feed PIFAs have been fully characterised and validated practically. These antennas, despite having very closely spaced feeds and not relying on spatial diversity techniques, are seen not to suffer from any detrimental performance effects – a high level of diversity gain is provided (close to the theoretical maximums, the measured correlation between the feeds is very low, a high level of channel capacity (spectral efficiency) is evident and about the centre working frequencies, the antennas are highly efficient.

The measurement procedures for multi-port antennas in the RC have been fully documented and all results have been benchmarked with simulated and/or theoretical results; excellent agreements are obtained. This fact proves the correctness and appropriateness of all procedures and parameters selected for the measurement campaigns.

For the conventional antenna arrays, a new equation has been developed which allows for the technique of de-embedding externally used power dividers and the calculation of the arrays' efficiency in one. It has been shown that this new developed technique is simple, easy to implement and accurate. This technique paves the way for further array characterisations in the RC, as the overall measurement techniques here have the potential to be faster, simpler and less uncertain than if other measurement facilities were to be employed.

A section has also been provided in this chapter on potential pitfalls to avoid during multi-port antenna measurements, as well as guidance on good practices to follow. It is recommended that these should always be applied.

References

[1] K. Rosengren and P. S. Kildal, 'Radiation efficiency, correlation, diversity gain and capacity of a six-monopole antenna array for a MIMO system: theory, simulation and measurement in reverberation chamber', *IEE Proceedings Microwaves Antennas and Propagation*, vol. 152, pp. 7–16, 2005.

[2] P. S. Kildal and K. Rosengren, 'Correlation and capacity of MIMO systems and mutual coupling, radiation efficiency, and diversity gain of their antennas, simulations and measurements in a reverberation chamber', *IEEE Communications Magazine*, vol. 42, pp. 104–112, December 2004.

[3] R. G. Vaughan and J. B. Andersen, 'Antenna diversity in mobile communications', *IEEE Transactions on Vehicular Technology*, vol. 36, pp. 149–172, 1987.

[4] P. Mattheijssen, M. H. A. J. Herben, G. Dolmans and L. Leyten, 'Antenna-pattern diversity versus space diversity for use at handhelds', *IEEE Transactions on Vehicular Technology*, vol. 53, pp. 1035–1042, 2004.

[5] B. H. Allen, M. Dohler, E. E. Okon, W. Q. Malik, A. K. Brown and D. J. Edwards (eds.), *Ultra-Wideband: Antennas and Propagation for Communications, Radar and Imaging*, Chichester: John Wiley & Sons, Ltd, 2007.

[6] H. T. Chattha, '*Planar Inverted F Antennas for Wireless Communications*', D Phil Thesis, Department of Electrical Engineering & Electronics, The University of Liverpool, 2010.

[7] H. T. Chattha, Y. Huang, S. J. Boyes and X. Zhu, 'Polarization and pattern diversity-based dual-feed planar inverted-F antenna', *IEEE Transactions on Antennas and Propagation*, vol. 60, pp. 1532–1539, 2012.

[8] A. Diallo, C. Luxey, P. Le Thuc, R. Staraj and G. Kossiavas, 'Study and reduction of the mutual coupling between two mobile phone PIFAs operating in the DCS1800 and UMTS bands', *IEEE Transactions on Antennas and Propagation*, vol. 54, pp. 3063–3074, 2006.

[9] A. Mak, C. R. Rowell and R. D. Murch, 'Isolation enhancement between two closely packed antennas', *IEEE Transactions on Antennas and Propagation*, vol. 56, pp. 3411–3419, 2008.

[10] C. Shin-Chang, W. Yu-Shin and C. Shyh-Jong, 'A decoupling technique for increasing the port isolation between two strongly coupled antennas', *IEEE Transactions on Antennas and Propagation*, vol. 56, pp. 3650–3658, 2008.

[11] W. Jakes, *Microwave Mobile Communications*, New York: John Wiley & Sons, Inc., 1974.

[12] P. S. Kildal, K. Rosengren, J. Byun and J. Lee, 'Definition of effective diversity gain and how to measure it in a reverberation chamber', *Microwave and Optical Technology Letters*, vol. 34, pp. 56–59, 2002.

[13] P. S. Kildal and K. Rosengren, 'Electromagnetic analysis of effective and apparent diversity gain of two parallel dipoles', *IEEE Antennas and Wireless Propagation Letters*, vol. 2, pp. 9–13, 2003.

[14] P. S. Kildal, *Foundations of Antennas: A Unified Approach*, Sweden: Studentlitteratur, 2000.

[15] X. Chen, 'Spatial correlation and ergodic capacity of MIMO channel in reverberation chamber', *International Journal of Antennas and Propagation*, pp. 1–7, 2012.

[16] E. Kreyszig, *Advanced Engineering Mathematics*, 8th ed.: Chichester: John Wiley & Sons, Ltd, 1999.

[17] J. Yang, S. Pivnenko, T. Laitinen, J. Carlsson and X. Chen, 'Measurements of diversity gain and radiation efficiency of the Eleven antenna by using different measurement techniques, *2010 Proceedings of the Fourth European Conference on Antennas and Propagation (EuCAP)*, IEEE, 1 April 2010, Barcelona, pp. 1–5.

[18] P. Hannan, 'The element-gain paradox for a phased-array antenna', *IEEE Transactions on Antennas and Propagation*, vol. 12, pp. 423–433, 1964.

[19] Y. W. Zhang and A. K. Brown, 'Octagonal ring antenna for a compact dual-polarized aperture array', *IEEE Transactions on Antennas and Propagation*, vol. 59, pp. 3927–3932, October 2011.

[20] H. A. Wheeler, 'Simple relations derived from a phased-array antenna made of an infinite current sheet', *IEEE Transactions on Antennas and Propagation*, vol. 13, pp. 506–514, 1965.

[21] A. K. Brown, 'Novel Broadband Antenna Arrays', ed. LAPC 2012: IET.tv, 2012.

7

Further Applications and Developments

Over the years, there have been many research groups around the world that have studied Reverberation Chambers (RCs) from theory to practice. Most notably, the group at NIST (the National Institute of Standards and Technology) in the United States has carried out a lot of fundamental research, developed measurement methods and explored applications. The RC as a measurement facility is still relatively young and there are many potential testing methods and applications to be explored and developed. In addition to the Electromagnetic Compatibility (EMC) (especially the radiated emission and immunity tests) and antenna measurements discussed in previous chapters, there have been many further applications and developments reported in the literature. The objective of this chapter is to introduce and discuss some of these applications and developments which have not been covered in the previous chapters. Our focus will be on shielding effectiveness measurements and antenna measurements.

7.1 Shielding Effectiveness Measurements

The Shielding Effectiveness (SE) is defined as the ratio of the signal received (from a transmitter) without the shield to the signal received inside the shield and can also be considered as the insertion loss when the shield is placed between the transmitting antenna and the receiving antenna. The shield or shielding enclosure is a structure (normally made of a conducting material) which can protect its interior from the effect

Reverberation Chambers: Theory and Applications to EMC and Antenna Measurements, First Edition.
Stephen J. Boyes and Yi Huang.
© 2016 John Wiley & Sons, Ltd. Published 2016 by John Wiley & Sons, Ltd.

of an interior electric or magnetic field or conversely, protects the surrounding environment from the effect of an interior electric or magnetic field. Mathematically the SE is defined as:

$$SE_{dB} = 10\log_{10}\left[\frac{Power\_without\_shielding}{Power\_with\_shielding}\right] \qquad (7.1)$$

where the SE is expressed in decibels (dB) since it is easy to deal with the wide dynamic range in practice. If the signal received is not in power, we may have:

$$SE_{dB} = 20\log_{10}\left[\frac{Voltage, current, field\_without\_shielding}{Voltage, current, field\_with\_shielding}\right] \qquad (7.2)$$

This is a very important parameter, especially for EMC applications. The shielding enclosure does not have to be a box, it can be a cable shielding structure or a ventilation window. There are some associated standards such as IEC 61000-5-7 [1] and IEC 61587-3 [2].

IEC 61000-5-7 [1] describes performance requirements, test methods and classification procedures for degrees of protection provided by empty enclosures against electromagnetic disturbances for frequencies between 10 kHz and 40 GHz. The shielding protection is measured for the purpose of demonstrating that the enclosure provides adequate shielding of electromagnetic energy to support acceptable performance of the complete assembled units when tested to applicable IEC standards. The purpose of this standard is to provide a repeatable means for evaluating the electromagnetic shielding performance of empty mechanical enclosures, including cabinets and subracks, and to specify a marking code to allow a manufacturer to select an enclosure with a known capability for attenuating electromagnetic fields. The requirements for immunity to various types of electromagnetic disturbances, including lightning and high-altitude electro-magnetic pulse (HEMP), will need to be considered by manufacturers when determining the need for application of this standard for specific equipment and the specific enclosure shielding requirements.

IEC 61587-3:2013 [2] specifies the tests for empty cabinets and subracks concerning electromagnetic shielding performance, in the frequency range of 30 MHz to 3 GHz. Stipulated attenuation values are chosen for the definition of the shielding performance level of cabinets and subracks for the IEC 60297 and IEC 60917 series. The shielding performance levels are chosen with respect to the requirements of the typical fields of industrial application. They will support the measures to achieve EMC, but cannot replace the final testing of compliance of the equipped enclosure. This second edition (2013) cancels and replaces the first edition issued in 2006. It constitutes a technical revision. The main technical changes with regard to the previous edition are that this edition has corrected the errors of electromagnetic code descriptions and the frequency range for the shielding performance is extended up to 3 GHz.

Another well-known standard in this subject is the IEEE Std 299.1 [3], which details how to measure the SE of enclosures and boxes having all dimensions between 0.1 and 2 m. It has included a number of different approaches.

Generally speaking, measuring the SE of small enclosures poses various problems. A main problem is associated with the internal resonances of the enclosure which occurs with the measuring of shielding effectiveness for any size enclosure, both physically large and small. Because of the resonant nature of the fields inside a shielding enclosure, the fields have an internal modal structure; as a result, measurement of the fields inside the enclosure is a function of the location where the measurement is performed. For large enclosures, two basic approaches can be used to overcome this problem. One approach is to sample the field at various locations in the enclosure and then take some type of average value of the power level inside the enclosure as suggested in Ref. [3]. However, this is not practical for physically small enclosures because moving a probe throughout the volume of a small enclosure would be problematic. The second approach is based on a nested RC technique (IEC 61000-5-7) [1], which is to place a smaller RC inside a larger RC. However, placing a probe (or antenna) in the centre of the small enclosure as is done in IEC 61000-5-7 poses difficulties. In addition, using conventional paddle mode stirring in a small enclosure would be problematic as well. That is, in most applications of measuring the SE of small enclosures it may not be possible to place a small mechanical stirring device inside the enclosures.

A frequency-stirred (averaging) RC approach may overcome these issues (see IEC 61587-3 [2] and [4] for details). This procedure assumes that the enclosure is physically small (<0.75 m in the linear dimension), but electrically large (which means the frequency should be high enough). A diagram of the set-up for the proposed approach is shown in Figure 7.1 where the small enclosure is placed inside an RC and the mechanical stirrer is not shown. This type of configuration is essentially a nested RC as discussed in IEC 61000-5-7. In this set-up, three antennas and a Vector Network Analyser (VNA) are used for the practical implementation. The source (e.g. Antenna 1 or 2 is connected to the VNA or an RF signal generator) is scanned over a given frequency range. Because some portion of the RF energy in the outer chamber will couple into the small enclosure, this causes frequency stirring of the RF energy in the small enclosure. As a result, all points in the small enclosure statistically have the same field levels for the data averaged over some bandwidth of frequencies [4]. Hence, the problem of sampling location is resolved, without the need to have a paddle (or stirrer) in the small enclosure. A hybrid approach combining mechanical stirring in the outer chamber with frequency stirring can be applied also as described in Ref. [5]. The beauty of using this method is that the signal passed through the shield may come from different incident angles, not necessary be normal to the surface which is a better reflection of the real world case.

To conduct the measurement, we use the set-up in Figure 7.1 and let Antenna 1 be the transmitting antenna. Antenna 2 and the small enclosure should be in the

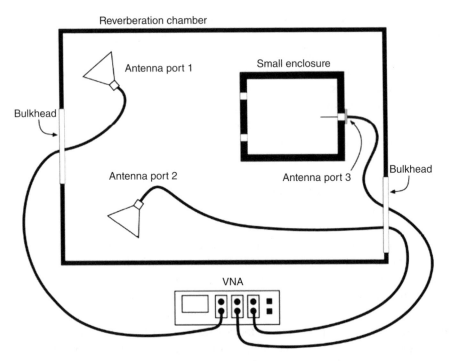

Figure 7.1   The set-up for the SE measurement using the basic frequency-stirred reverberation chamber technique. Source: IEEE Std 299.1-2013 [3]. Reproduced with permission of IEEE.

Equipment under Test (EUT) area of the RC. The aim is to measure the power without shielding and the power with shielding, thus Equation 7.1 can be used to calculate the SE. These two powers can be measured at the same time since, statistically, the power received by Antenna 2 should be the power without shielding, while the power received by Antenna 3 should be the power with shielding if the chamber is well stirring either mechanically or in frequency or both. It is important to point out that the orientation and location of Antenna 2 may affect the measurement results. The small enclosure should not block Antenna 2 in the measurement. Also, since the antennas are unlikely perfectly matched with the feedlines, especially over a frequency band, we have to take the impedance mismatch into account when Equation 7.1 is applied, thus the SE can be expressed using the measured S-parameters as:

$$\mathrm{SE}_{\mathrm{dB}} = 10\log_{10}\left[\frac{\left\langle\left|S_{31}\right|^2\right\rangle \Big/ \left(1-\left\langle\left|S_{33}\right|^2\right\rangle\right)}{\left\langle\left|S_{21}\right|^2\right\rangle \Big/ \left(1-\left\langle\left|S_{22}\right|^2\right\rangle\right)}\right] \tag{7.3}$$

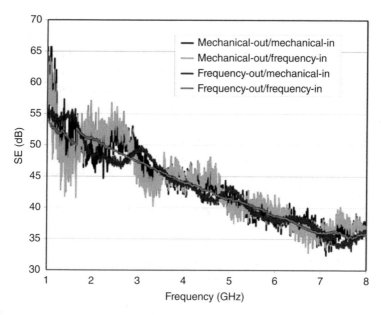

Figure 7.2 SE obtained from four different reverberation chamber approaches for an aperture consisting of grid of holes. Source: Greco and Sarto [5]. Reproduced with permission of IEEE.

where the $\langle \rangle$ represents the ensemble average, averaged over some frequency bandwidth (and paddle position if a combination of frequency and mechanical stirring is used in the outer RC):

$$\left\langle \left| S_{ab} \right|^2 \right\rangle = \frac{1}{N} \sum_{i=1}^{N} \left| S_{ab} \left(x_i \right) \right|^2 \tag{7.4}$$

N is the total number of frequency and/or paddle points used for this averaging approach; a and b are integers 1, 2 or 3; and x is a variable which could be the frequency or stirrer position, depending on the stirring method used.

Although in Figure 7.1 a multi-channel VNA is shown, but in practice we can just use a typical 2-port VNA, which means we need to measure the signals at Port 2 and Port 3 at two different times, thus the total measurement time is doubled (but a less expensive VNA is used as a trade-off). An example of measured SE of an aperture consisting of grid holes from Ref. [5] is given in Figure 7.2. Four different RC approaches were employed and they corresponded to different combinations of frequency stirring and mechanical stirring (see figure legend). From this comparison, we can see that all four approaches give similar results for the SE of the small enclosure. Higher SE values were obtained for lower frequencies. The measurement accuracy is not just determined by the measurement method, but also by other aspects, such as the connection and cables.

Here we have presented the standard SE measurement approach. It is interesting to note how the SE measurement techniques have evolved. For example, in an earlier research conducted in 1988 [6], the power levels in a small enclosure were monitored by a small monopole probe (or antenna) placed on one of the interior walls of the small enclosure. It was shown that the normal component of the electric field at the surface of a wall in a well-stirred cavity had the same statistics as a probe placed anywhere in the cavity. Thus, as long as the small enclosure was well stirred (i.e. through frequency-stirring), a small monopole probe placed on the inside wall would give the same average power level inside the small enclosure as that of an antenna placed in the centre of the small enclosure, which means that Antenna 3 in Figure 7.1 can be at any place inside the enclosure and doesn't have to be a monopole. It was concluded that the frequency stirring in the outer chamber could be done with or without a conventional mechanical stirring process in the large outer RC. However, using a combination of both the frequency and mechanical stirring in the large outer RC could improve the accuracy in the measurements. These results are now validated and widely accepted, and some have been adopted in various standards.

The SE measurement seems to be a relatively well-studied subject but there are still continuous efforts on developing new and hopefully more efficient and accurate techniques and methods. For example, recently a method to determine the SE in an RC using radar cross-section simulations with a planar wave excitation was reported [7] which may not be practical for application and may even contain errors (a negative SE was obtained in the paper due to the use of a different/ wrong definition of the SE), but it is an interesting angle to see how the SE might be obtained.

Another example is that a new method was proposed to determine the SE using just one antenna [8]. Traditionally, the SE measurement requires at least two (in the anechoic chamber (AC)) or three antennas (in the RC as shown in Figure 7.1). In this paper, only one antenna was needed in the whole measurement procedure and there was no need for a reference antenna. The main idea is that the SE can be measured quickly by comparing the Q factors or the decay times with and without the shielding. Measurements had been conducted to evaluate and verify the proposed method by comparing the results with those from the nested RC method. Good agreements were obtained. The main limitation of this method is that it is only suitable for the SE, is not too large, typically less than 30 dB, otherwise the measurement would not be accurate.

The discussion in this section has been limited to enclosures. In practice, the SE of some componets such as shielding materials, connectors and air vents is to be measured. The same methods can be applied to a dual RC if a nested RC method is not suitable. The device under test can be placed between the chambers. Some commercial products from such as EST-Lingren are available on the market.

7.2 Antenna Radiation Efficiency Measurements without a Reference Antenna

In Chapter 5, we have already defined antenna radiation efficiency and introduced the measurement method in an RC where a reference antenna with known antenna efficiency is required. But in practice this requirement could be an issue: an antenna with known efficiency over the desired frequency band might not be available or the actual value of the efficiency might have changed since last calibration. Thus an alternative method of measuring antenna efficiency in an RC is needed.

The group at NIST recently has found a solution which was published [9] in 2012. Three different approaches for determining both the radiation and total efficiencies of an unknown antenna were presented. They were one-antenna, two-antenna approach and three-antenna approaches. The basic idea behind these approaches is to find the time domain response of the RC and use the Q factors obtained from the time domain (Q_{TD}) and the frequency domain (Q_{FD}) to calculate the radiation efficiency of the antennas without using a reference antenna [9].

For the one-antenna approach, the measurement set-up is shown in Figure 7.3 and we only need to use the Antenna under Test (AUT) and one port of the VNA. Other antennas are not needed, thus it is simpler. It was shown in Ref. [9] that the total efficiency of the AUT can be computed using

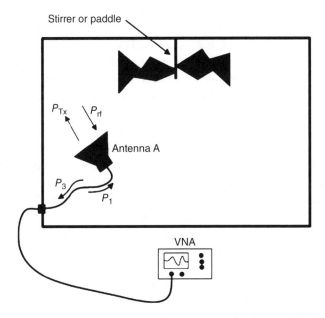

Figure 7.3 The measurement set-up for one-antenna approach using the RC. Source: Holloway et al. [9]. Reproduced with permission of IEEE.

$$\eta_{\text{Total}} = \sqrt{\frac{C_{\text{RC}}}{2\omega} \frac{\left\langle |S_{11_s}|^2 \right\rangle}{\tau_{\text{RC}}}} \tag{7.5}$$

where ω is the angular frequency; $C_{\text{RC}} = \dfrac{16\pi^2 V}{\lambda^3}$ is the chamber constant and V is the volume of the chamber; τ_{RC} is the chamber time constant ($=Q/\omega$, Q is the chamber quality factor) and $\left\langle |S_{11_s}|^2 \right\rangle$ represents the stirred energy contribution of the AUT $S11$ inside the RC.

Thus a measurement of $\left\langle |S_{11_s}|^2 \right\rangle$ and τ_{RC} is all that is required to obtain the total radiation efficiency of a single antenna. If radiation efficiency is required, it can be obtained using

$$\eta = \frac{\eta_{\text{Total}}}{1 - |S_{11}|^2} = \sqrt{\frac{C_{\text{RC}}}{2\omega} \frac{\left\langle |S_{11_s}|^2 \right\rangle}{\tau_{\text{RC}}}} \Big/ \left(1 - |S_{11}|^2\right) \tag{7.6}$$

The measurement procedures for two-antenna approach and three-antenna approach are very similar. They require the measurements of $\left\langle |S_{11_s}|^2 \right\rangle$ and τ_{RC} as well. The main differences are in the assumptions on the relationship between the power reflected into the antenna and the power received [9]. The three-antenna approach is more general and does not require such assumptions. The measurement results are very similar for all three approaches as we can see from the example in Figure 7.4 (the largest difference is <5%). The two-antenna approach seems to give the smallest uncertainty.

Compared with the RC method, using a reference antenna, the main advantage of this new method is that it does not require a reference antenna with known radiation efficiency. But one has to calibrate the chamber carefully in the time domain to obtain the time constant – this is a time-consuming exercise. The good thing is that one just needs to calibrate the chamber once if the chamber performance/structure/loading is not changed, and the data can be used repeatedly.

It should be pointed out that, for these three approaches, a pre-condition has to be satisfied: the overall loss is dominated by the chamber (i.e. the antenna cannot be very lossy). If the antenna is very lossy (radiation efficiency is very small which is true for such as implant antennas), the time domain response obtained is not dominated by the chamber, and the methods proposed in Ref. [9] may not work.

To overcome this limitation, a modified two-antenna method has been proposed [10] and the measurement set-up is shown in Figure 7.5. By combing the conventional

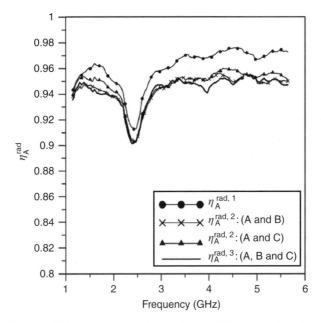

Figure 7.4 Radiation efficiency: a comparison of three different approaches for Antenna A in Ref. [9]. Source: Holloway et al. [9]. Reproduced with permission of IEEE.

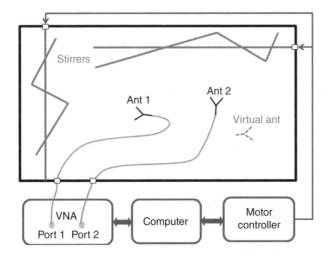

Figure 7.5 The measurement set-up for the modified two-antenna method.

reference antenna method and the one-antenna method proposed in Ref. [9], and introducing a virtual antenna, the proposed method does not need a reference antenna and can be applied to highly lossy antennas. That is, the one-antenna method is first applied using a high efficiency antenna, thus the radiation efficiency of this antenna is obtained and the chamber transfer function can also be obtained by taking advantage of the enhanced backscatter effect, and then the radiation efficiency of the AUT can be determined by using the traditional reference antenna method and the obtained radiation efficiency. A virtual antenna is involved to complete the whole procedure (like a reference antenna in the conventional reference antenna method). Measurements are conducted to verify the effectiveness of the proposed method. It can be seen from Figure 7.6 that the result from this modified method is almost the same as the ones obtained from the one-antenna and two-antenna methods in Ref. [9] for a very good efficiency antenna in Figure 7.6a but they are very different for the case of a very lossy antenna in Figure 7.6b. This method can be regarded as a generalised form of the two-antenna method in Ref. [9]. A main

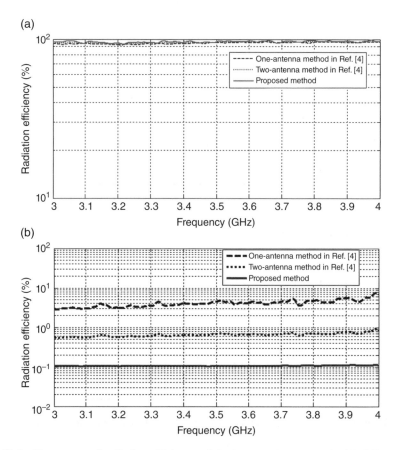

Figure 7.6   The measured radiation efficiency of the Antenna 1 connected with different attenuators. (a) 0 dB attenuation and (b) 30 dB attenuation [10].

condition for this modified method is that one of the two antennas has to be efficient, which should not be a problem in practice.

7.3 Antenna Diversity Gain Measurements without a Reference Antenna

The diversity gain is the most important parameter for diversity and Multiple Input and Multiple Output (MIMO) antennas. It can be measured using an RC as discussed in Chapter 6. In addition to the AUT, the traditional method for this type of measurement needs to use two additional antennas: a transmitting antenna and a reference antenna with known radiation efficiency. Again this could be an issue in practice.

Thus a new method was proposed recently in Ref. [11] to measure the diversity gain of a diversity/MIMO antenna system using an RC. In the proposed method, one branch of the AUT is used as the reference antenna, thus the requirement for the standard reference antenna can be removed and only a transmitting antenna is required. The results obtained from the new and traditional methods were compared, and a very good agreement has been achieved.

Further research has been conducted recently to further simplify the measurement system by using just the AUT only for the whole measurement [12], thus both the transmitting and reference antennas are not required! This is realised using the one-antenna or two-antenna method for the radiation efficiency measurement in an RC and a virtual antenna using the enhanced backscattering effect [9]. A two-port planar inverted-L antenna was used as an example to evaluate the proposed method. The obtained Cumulative Distribution Function (CDF) plots for both branches and the combined signal are given in Figure 7.7, and the apparent diversity gain was about 10.19 dB which is in good agreement with other established methods.

However, there are some conditions and limitations on this new method: it is based on the one-antenna or two-antenna method introduced in Ref. [9], thus the chamber must be well-stirred and the losses in the chamber are dominated by the chamber wall losses (including any load) as discussed in Section 7.2. Another condition is the mutual coupling between antenna elements must be very small to ensure the accuracy of this measurement method. Both conditions should be fine for most diversity and MIMO antennas. The proposed method has reduced the system requirement and measurement time without compromising the measurement accuracy, and is therefore a very attractive method for diversity and MIMO antenna measurements. More details can be found from Ref. [12].

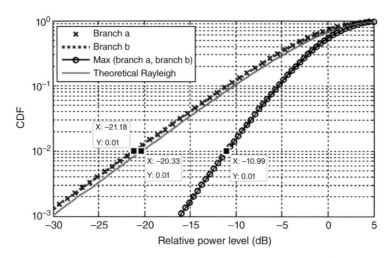

Figure 7.7 CDF plots of both branches and the combined signal using the proposed method in Ref. [12].

7.4 Wireless Device and System Evaluation

The wireless industry has been growing for the last 20 years or so. Many wireless devices and systems have been developed and have entered into our daily life. This trend is set to continue. Most of these devices and systems are developed for the multipath environment, and a real site test and evaluation could be inclusive and not repeatable. However the RC is a well-controlled multi-path environment and can simulate a wide range of the radio propagation channels [13–17], there is a growing interest in using the RC to evaluate wireless device and system performance [18, 19].

In Ref. [15], the use of the RC to simulate fixed wireless propagation environments including effects such as narrowband fading and Doppler spread have been demonstrated. These effects have a strong impact on the quality of the wireless channel and the ability of a receiver to decode a digitally modulated signal. Different channel characteristics such as power delay profile and Root Mean Square (RMS) delay spread are varied inside the chamber by incorporating various amounts of absorbing material. In order to illustrate the impact of the chamber configuration on the quality of a wireless communication channel, Bit Error Rate (BER) measurements were performed inside the RC for different loadings, symbol rates and paddle speeds. Measured results acquired inside a chamber were compared with those obtained both in an actual industrial environment and in an office. Figure 7.8 shows the measured Power Delay Profiles (PDP) in an RC and an oil refinery in the United States. It can be seen that the PDP from the oil refinery environment best matches the RC with three absorbers inside.

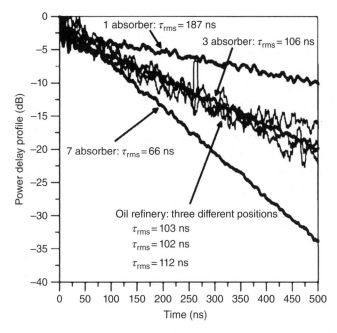

Figure 7.8 Comparison of the power delay profile measured in an RC for different amounts of loading to the power delay profile measured in an oil refinery in the United States. Source: Genender et al. [15]. Reproduced with the permission of IEEE.

Thus RCs are emerging as a test facility for testing wireless devices and for emulating different wireless multi-path environments. The commonly used quantities for characterising the chambers in wireless applications are:

1. the chamber quality factor Q,
2. the chamber decay time (τ_{RC}),
3. the RMS delay spread of the time domain chamber response τ_{rms} and
4. the coherence bandwidth BW of the frequency domain transfer function of the chamber.

Analytic expressions that relate τ_{RC} and BW and the relationship between τ_{rms} and BW have been detailed in Chapter 2 and are also available in the literature. However, these expressions neglect the early-time behaviour of the chamber (the time before a chamber reaches a reverberant condition), and hence can give inconsistent results when one is analysing experimental data. In Ref. [17], the relationship between BW, τ_{RC} and τ_{rms} for realistic chamber behaviours was discussed, and the expressions for these relationships were given by taking the early-time behaviour of the RC into account. This early-time behaviour is crucial when one tries to assess and compare these different quantities in experimental data. It was shown that the relationship between these quantities could be different for different chambers (i.e. different chamber sizes and loading conditions). The model presented in Ref. [17] illustrated how the early-time behaviour could affect these chamber characteristic quantities for loaded and unloaded chambers.

It has been demonstrated that it is possible to simulate various types of real-world wireless propagation environments inside an RC by various loading and even using channel emulator. The RC may be used as a reliable, repeatable test environment and, in contrast to an AC or free space, a very efficient testing facility for these multi-path wireless propagation measurements.

It is interesting to see that some wireless systems have been evaluated using an RC. For example, in Ref. [18], a Wireless Local Area Network (WLAN) with MIMO, which is very similar to our measurement in Chapter 4 as shown in Figure 4.9, was tested inside a large RC not only for the immunity evaluation, but also for the system performance. A whole link established between commercial devices operated inside the RC under different loading conditions, varying the stirrer rotating speed and the paddle dimensions. The effects of this multi-path environment on system performance were assessed by measuring the number of cyclic redundancy check error and the number of data retries. The comparison between the two systems was done adopting the same chamber loading condition and also stressing the system reducing the load, up to the link connection limit. Results have demonstrated better performance for the MIMO system, which was able to maintain a 6 Mb/s data rate connection with a chamber quality factor of 21 000.

We believe that more RC applications will be seen in the wireless device and system evaluations in the coming years and associated standards will be developed. The channel effects will include not only the attenuation (PDP), delay spread, but also others such as the Doppler frequency shift. The combination of the RC and advanced wireless measurement equipment will be required for some applications.

7.5 Other Reverberation Chambers and the Future

There have been many other interesting studies in the RC over the years. Here are just some examples.

7.5.1 Reverberation Chamber Shapes

At the moment, almost all RCs are rectangular in shape, although this is not necessarily the best shape in application. In practice, the triangular tent is more popular than the rectangular one due to its stability and simplicity. The modes inside triangular chambers have been studied in Ref. [19] which showed that, using the same amount of material to construct a chamber, the more asymmetric it is, the more possible modes may exist inside such a chamber. Thus a symmetrical rectangular chamber may be not an optimised shape for the RC.

7.5.2   Flexible Reverberation Chamber

One of the problems in current RCs is that the chamber is fixed, not transportable, but sometimes measurements and tests are required for installed systems which cannot be moved into a fixed RC, thus a transportable RC would be ideal for such an application. Some initial study was reported by, for example [20], where conducting cloth (flexible material) was used to construct the chamber (still rectangular in shape and it would be easier to make a triangular one), no mechanical stirrer was used – this is an advantage to maximise the use of the available space. Some interesting results were obtained.

Thinking along a similar line, Gruden et al. [21] discussed how to convert the existing room into an RC with removable mode stirrier for mobile phone antenna tests.

7.5.3   Millimetre Wave Reverberation Chambers

Reverberation chambers have the lowest usable frequency. But what happens at very high frequencies, such as into millimetre waves? In principle, the conventional RC should work as well. The main change could be the loss since the skin-depth is smaller in millimetre waves than in RF and microwave. Also the wavelength is much smaller, therefore the chamber can be much smaller. An initial study was reported in Ref. [22] and showed that the Q factor did not increase much in millimetre and the chamber loss had little change in simulation which was confirmed by measurement over a small bandwidth. It would be good to see a more comprehensive study into a wider range of the chamber characteristics in millimetre range.

7.5.4   Future Directions

Although there have been a lot of research into the RC, there are still some hard questions not yet answered, such as for a given chamber, how to design an optimised stirrer to obtain the largest EUT area and the lowest usable frequency? We believe that the research into the RC will have at least two related directions.

One is from the chamber design point of view: more effort will be put into the optimised RC designs, including the stirring methods. The source stirrer method is a very attractive concept, but hard to implement in practice using the current knowledge.

Another is from the application point of view: more applications of the RC will be explored. We believe that the application in wireless device and system evaluation is going to grow, so is the application into millimetre waves.

7.6 Summary

In this chapter we have discussed the measurement of the shielding effectiveness of enclosures and materials. We have also discussed some of the latest developments in antenna efficiency measurements and diversity/MIMO antenna measurements. Some interesting ideas about RCs were briefly introduced and it was predicted that more applications of the RC would appear and be applied especially in wireless device and system evaluation.

References

[1]   IEC 61000 Part 5-7: Installation and mitigation guidelines – Degrees of protection provided by enclosures against electromagnetic disturbances. 2001.

[2]   IEC 61587-3: Mechanical structures for electronic equipment – Tests for IEC 60917 and IEC 60297 – Part 3: Electromagnetic shielding performance tests for cabinets and subracks. 2013.

[3]   IEEE Std 299.1-2013: IEEE standard method for measuring the shielding effectiveness of enclosures and boxes having all dimensions between 0.1 and 2 m. IEEE, October 2013.

[4]   C. L. Holloway, D. A. Hill, M. Sandroni, J. M. Ladbury, J. Coder, G. Koepke, A. C. Marvin and Y. He, 'Use of reverberation chambers to determine the shielding effectiveness of physically small, electrically large enclosures and cavities', *IEEE Transactions on Electromagnetic Compatibility*, vol. 50, pp. 770–782, 2008.

[5]   S. Greco and M. S. Sarto, 'New hybrid mode-stirring technique for SE measurement of enclosures using reverberation chambers', *Proceedings of IEEE Electromagnetic Compatibility International Symposium*, 9–13 July 2007, Honolulu, HI, pp. 1–6.

[6]   M. O. Hatfield, 'Shielding effectiveness measurements using mode-stirred chambers: a comparison of two approaches', *IEEE Transactions on Electromagnetic Compatibility*, vol. 30, pp. 229–238, 1988.

[7]   R. J. Long and A. S. Agili, 'A method to determine shielding effectiveness in a reverberation chamber using radar cross-section simulations with a planar wave excitation', *IEEE Transactions on Electromagnetic Compatibility*, vol. 56, pp. 1053–1060, 2014.

[8]   Q. Xu, Y. Huang, *et al.*, 'Shielding effectiveness measurement of an electrically large enclosure using one antenna', *IEEE Transaction on Electromagnetic Compatibility* in press.

[9]   C. L. Holloway, H. A. Shah, R. J. Pirkl, W. F. Young, D. A. Hill and J. Ladbury, 'Reverberation chamber techniques for determining the radiation and total efficiency of antennas', *IEEE Transactions Antennas Propagation*, vol. 60, no. 4, pp. 1758–1770, 2012.

[10] Q. Xu, Y. Huang, X. Zhu, L. Xin, *et al.*, 'A modified two-antenna method to measure the radiation efficiency of antennas in a reverberation chamber', *IEEE Antennas and Wireless Propagations Letters*, 2015, doi:10.1109/LAWP.2015.2443987.

[11] Q. Xu, Y. Huang, X. Zhu, L. Xin, *et al.*, 'A new antenna diversity gain measurement method using a reverberation chamber', *IEEE Antennas and Wireless Propagation Letters*, 2015, vol. 14, pp. 935–938.

[12] Q. Xu, Y. Huang, X. Zhu, L. Xin, *et al.*, 'Diversity gain measurement a reverberation chamber without extra antennas', *IEEE Antennas and Wireless Propagation Letters*, 2015, doi:10.1109/LAWP.2015.2417655.

[13] C. L. Holloway, D. A. Hill, J. M. Ladbury, P. Wilson, G. Koepke and J. Coder, 'On the use of reverberation chambers to simulate a controllable Rician radio environment for the testing of wireless devices', *IEEE Transactions on Antennas and Propagation, Special Issue on Wireless Communications*, vol. 54, no. 11, pp. 3167–3177, 2006.

[14] H. Fielitz, K. Remley, C. Holloway, Q. Zhang, Q. Wu and D. Matolak, 'Reverberation-chamber test environment for outdoor urban wireless propagation studies', *IEEE Antennas Wireless Propagation Letters*, vol. 9, pp. 52–56, 2010.

[15] E. Genender, C. L. Holloway, K. A. Remley, J. M. Ladbury and G. Koepke, 'Simulating the multipath channel with a reverberation chamber: application to bit error rate measurements', *IEEE Transactions on Electromagnetic Compatibility*, vol. 52, no. 4, pp.766–777, 2010.

[16] P.-S. Kildal, C. Orlenius and J. Carlsson, 'OTA testing in multipath of antennas and wireless devices with MIMO and OFDM', *Proceedings of the IEEE*, vol. 100, no. 7, pp. 2145–2157, 2012.

[17] C. L. Holloway, H. A. Shah, R. J. Pirkl, K. A. Remley, D. A. Hill and J. Ladbury, 'Early time behavior in reverberation chambers and its effect on the relationships between coherence bandwidth, chamber decay time, RMS delay spread, and the chamber buildup time', *IEEE Transactions on Electromagnetic Compatibility*, vol. 54, no. 4, pp.717–725, 2012.

[18] R. Recanatini, F. Moglie and A. M. Primiani, 'Performance and immunity evaluation of complete WLAN systems in a large reverberation chamber', *IEEE Transactions on Electromagnetic Compatibility*, vol. 55, no. 5, pp.806–815, 2013.

[19] Y. Huang, 'Triangular screened chambers for EMC tests', *Measurement Science and Technology*, vol. 10, pp. 121–124, 1999.

[20] F. Leferink, 'In-situ high field strength testing using a transportable reverberation chamber', *19th International Zurich Symposium on EMC*, 19–22 May 2008, Singapore.

[21] M. Gruden, P. Hallbjorner and A. Rydberg, 'Large *Ad Hoc* shield room with removable mode stirrer for mobile phone antenna tests', *IEEE Transactions on Electromagnetic Compatibility*, vol. 55, no. 1, pp. 21–27, 2013.

[22] A. K. Fall, P. Besnier, C. Lemoine and R. Sauleau, 'Design and experimental validation of a mode-stirred reverberation chamber at millimeter waves', *IEEE Transactions on Electromagnetic Compatibility*, 2014, doi:10.1109/TEMC.2014.2356712.

Appendix A

Deduction of Independent Samples

The purpose of this appendix section is to detail the complete procedure as to how the Non-Line of Sight Number (NLoS) of independent samples can be calculated in Reverberating Chamber (RC) measurements. This technique is applied when deducing this quantity for use in the theoretical uncertainty formula and also to determine the stirrer efficiency.

To calculate the NLoS number of independent samples, the use of the autocorrelation function is considered. The term 'autocorrelation' can be defined as the correlation of a signal with itself.

If each sample is a linear power quantity, the offset delta (Δ) at which the autocorrelation has dropped to a given value can be found and the number of independent samples (N_{IND}) determined from:

$$N_{IND} = \frac{N_{MEASURED}}{\Delta} \tag{A1.1}$$

where $N_{MEASURED}$ is simply the total number of measurement samples.

For one specific measured frequency value and assuming 718 measured samples, all the samples obtained from various stirrer increments for a full revolution should be obtained and collated in a one-column vector. The column vector is permuted such that the last value in the original vector is placed as the first value in a newly created vector

Reverberation Chambers: Theory and Applications to EMC and Antenna Measurements, First Edition.
Stephen J. Boyes and Yi Huang.
© 2016 John Wiley & Sons, Ltd. Published 2016 by John Wiley & Sons, Ltd.

(i.e. vector two). This permuting action continues such that the penultimate value in the original vector is placed first in a newly created third vector and so on until a 718×718 (assuming one degree mechanical stirring intervals and polarisation stirring) matrix exists; in essence all the values in the original matrix had been permuted to form new vectors in their own right.

This permuting action is detailed as follows:

$$x_1, x_2, x_3, x_4 \ldots x_{N-1}, x_N$$
$$x_N, x_1, x_{2,} x_3, x_4 \ldots x_{N-1}$$
$$x_{N-1}, x_N, x_1, x_2, x_3, x_4 \ldots x_{N-2}$$

And so on, until all samples have been permuted.

Once the permuting of the vectors has been completed, the quantity of delta needs to be found. The quantity delta is a function of the total number of measurement samples and should not necessarily be approximated by the $1/e = 0.37$ criterion. This 0.37 criterion is only valid assuming 450 measured samples only [1]. To calculate delta, the theoretical probability density function (PDF) form of the correlation coefficient (ψ) can be used, as covered in Refs. [1] and [2]. In (A2), p must be estimated accordingly.

$$\psi(r) = \frac{N-2}{\sqrt{2\pi}} \times \frac{\Gamma(N-1)}{\Gamma(N-0.5)} \times \frac{\left(1-p^2\right)^{(N-1)/2}\left(1-r^2\right)^{(N-4)/2}}{\left(1-pr\right)^{N-3/2}} \times \left[1 + \frac{1+pr}{4(2N-1)} + \cdots\right]$$

$$\text{(A1.2)}$$

where p = expected value of r, Γ is the gamma function, N is the total number of measurement samples and r can be found by [3]

$$r = \frac{\dfrac{1}{N}\sum_1^n \left(A_{mn} - \bar{A}\right)\left(B_{mn} - \bar{B}\right)}{\sqrt{\left(\left(\sum_1^n \left(A_{mn} - \bar{A}\right)^2 / N - 1\right)\left(\sum_1^n \left(B_{mn} - \bar{B}\right)^2 / N - 1\right)\right)}} \qquad \text{(A1.3)}$$

where A = original (unpermuted) vector, B = each permuted vector in turn, mn stand for the row and column vectors and the bar terms above A and B signify the mean.

The PDF function of Equation A2 as depicted in Figure A.1 will be evident. In this case, Figure A.1 has been plotted for 718 samples assuming a value of $p = 0.4601$.

The next step concerns the deduction of the probability of occurrence (α) and the formulation of the null hypothesis. Alpha is determined by integrating the PDF form of

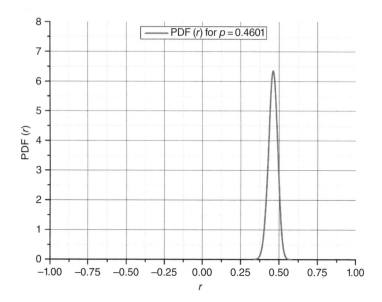

Figure A.1   PDF form of autocorrelation function.

Figure A.1 to determine the probability of occurrence of (r), which is smaller than or equal to the critical value (p_O). Thus:

$$\alpha = \int_{-1}^{p_O} \psi(r)\,dr \tag{A1.4}$$

The theoretical PDF is integrated until alpha reaches a given confidence interval (usually 0.05 (95% confidence) or 0.01 (99% confidence)). Once that value is reached, it will correspond to a given correlation value. This value must then be tested and either accepted or rejected according to the null hypothesis set out in [1]; thus:

If $r \leq p_O$ the hypothesis must be rejected. If this is the case, the probability that the sample originates from the assumed population is $\leq \alpha$ and p must be re-chosen [1]. Once an acceptable value has been found, it is applied as in Figures A.2 and A.3 to determine delta and the number of independent samples found from Equation A.1. For 718 samples, p has been determined as 0.4601 and this yields a critical value p_O of 0.394 at $\alpha = 0.01$.

Enlarging Figure A.3 about the point where the correlation will have dropped to the critical value of 0.394 (deduced for 718 measured samples), the value of delta can be found as shown in Figure A.3.

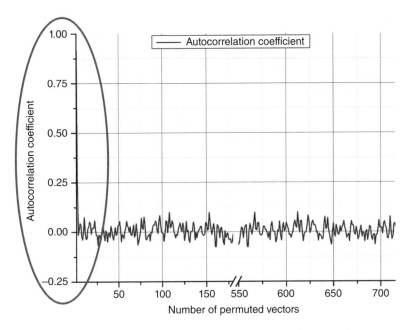

Figure A.2   Autocorrelation as a function of number of permuted vectors.

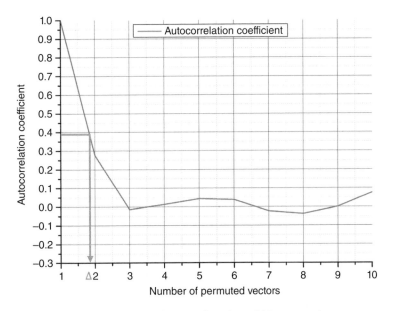

Figure A.3   Autocorrelation as a function of 10 permuted vectors.

References

[1] H. G. Krauthauser, T. Winzerling, and J. Nitsch, 'Statistical interpretation of autocorrelation coefficients for fields in mode-stiffed chambers', *2005 International Symposium on Electromagnetic Compatibility*, 2005. EMC 2005, Vol. **2**, 12 August 2005, Chicago, IL, pp. 550, 555.

[2] J. R. Taylor, *An Introduction to Error Analysis: The Study of Uncertainties in Physical Measurements*, 2nd ed.: Sauselito: University Science Books, 1997.

[3] 'BS EN 61000-4-21:2011 Electromagnetic compatibility (EMC). Testing and measurement techniques. Reverberation chamber test methods,' ed, 2011.

Appendix B

Multivariate Normality Test for SIMO Channels

The purpose of this appendix section is to statistically test the measured single input multiple output (SIMO) channel samples to determine whether or not they can be reasonably assumed to come from a normal distribution. As previously stated in Chapter 6, if the samples can be confirmed to be normally distributed, it would eliminate a potential source of uncertainty with regards to overestimating the levels of channel capacity, and further reliance can be placed on the measured/calculated values.

Similar to the statistical recommendations in Chapter 2, the Lilliefors test has been applied, and this evidence is compared to scatter plots to re-enforce any conclusions drawn from this analysis. In the Lilliefors test, the null hypothesis has been configured to assess whether the measured SIMO channel samples can be expected to come from a normal distribution and the confidence interval selected in the test is 95%. The test effectively measures the maximum distance from the deduced empirical cumulative distribution function (CDF) from the measured data and compares to a theoretical normal distribution having the same mean and standard deviation.

Figure B.1 details the Lilliefors test results for the co-polarised PIFA (feed 1) and Figure B.2 charts the scatter plot data for feed 1 to re-enforce the evidence from the Lilliefors test.

From Figure B.1 it can be seen that the Lilliefors test has accepted the hypothesis that the samples do come from a normal distribution at a 95% confidence interval.

Reverberation Chambers: Theory and Applications to EMC and Antenna Measurements, First Edition.
Stephen J. Boyes and Yi Huang.
© 2016 John Wiley & Sons, Ltd. Published 2016 by John Wiley & Sons, Ltd.

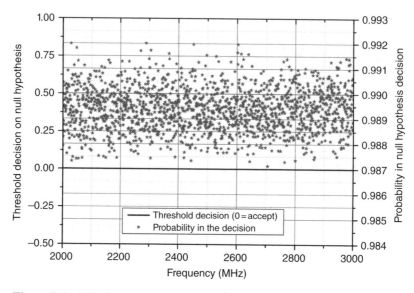

Figure B.1   Lilliefors test decision on normality for co-polarised PIFA feed 1.

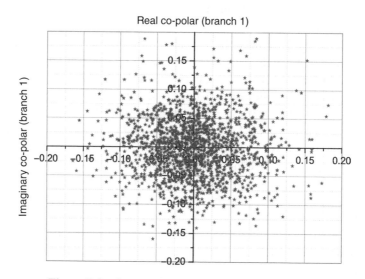

Figure B.2   Scatter plot for co-polarised PIFA feed 1.

From Figure B.2 it can be seen that the vast majority of samples are centred around the (0,0) axis providing further evidence to re-enforce the Lilliefors test's decision. Figures B.3 and B.4 detail the evidence obtained for the co-polarised PIFA feed 2.

Again, evidence is returned that suggests with a degree of confidence that the measured samples for feed 2 are also normally distributed. Moving on, Figures B.5 and B.6 detail the evidence obtained for the cross-polarised PIFA feed 1.

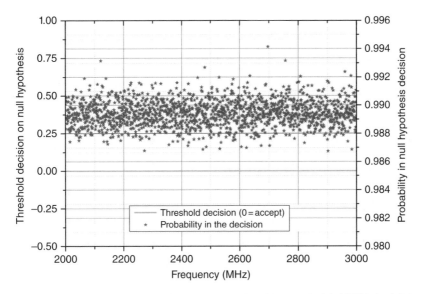

Figure B.3   Lilliefors test decision on normality for co-polarised PIFA feed 2.

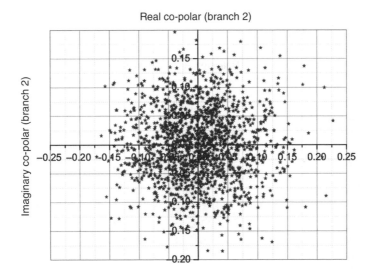

Figure B.4   Scatter plot for co-polarised PIFA feed 2.

Likewise, the evidence obtained for the cross-polarised PIFA feed 2 is depicted in Figures B.7 and B.8.

From the evidence issued in this section it is possible to conclude with a reasonable degree of certainty that the measured channel samples for both PIFAs used in the diversity gain and channel capacity measurements are normally distributed. With regards to the channel capacity calculations, it also serves to provide confidence that the result is not liable to overestimation as a result of non-normality.

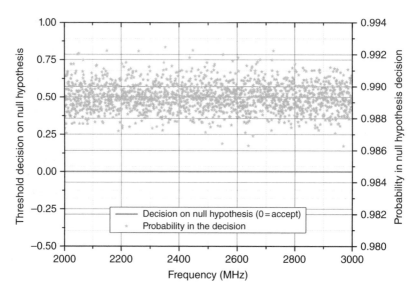

Figure B.5   Lilliefors test decision on normality for cross-polarised PIFA feed 1.

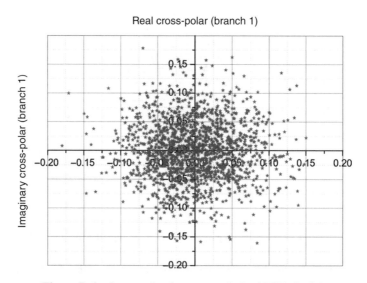

Figure B.6   Scatter plot for cross-polarised PIFA feed 1.

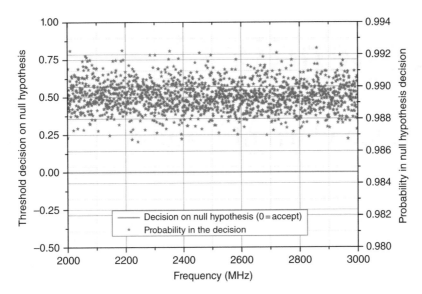

Figure B.7 Lilliefors test decision on normality for cross-polarised PIFA feed 2.

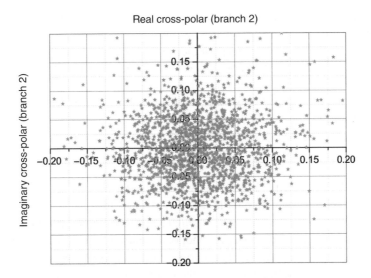

Figure B.8 Scatter plot for cross-polarised PIFA feed 2.

Appendix C

Surface Current Nature

The purpose of this appendix section is to depict the nature of the induced surface current with respect to the design of mechanical stirring paddles. This section provides the following evidence:

1. It will depict the specific size of the cuts interacting with a larger wavelength plane wave more effectively than standard designs.
2. It will clearly illustrate the concept of increasing the current path length and increasing the resonant capability – both crucial to the overall design strategy.
3. It clearly shows the manner as to how broadband performance is provided – that is, different cuts on the paddle surface interact with different-sized plane waves to allow for a more efficient operation.
4. Proof is provided that show points (1), (2) and (3) are achieved considering two orthogonal polarisations.

Hence, evidence to further re-enforce why the design strategy of implementing cuts on the paddle surfaces works better at lower modal densities.

The induced surface current has been calculated from a plane wave excitation. The excitation has been configured in two orthogonal polarisations from a direction normal to the plate surface. It is accepted that in operation in the RC, this incoming direction will come from many different angles as the paddles rotate; however, the purpose here

Reverberation Chambers: Theory and Applications to EMC and Antenna Measurements, First Edition.
Stephen J. Boyes and Yi Huang.
© 2016 John Wiley & Sons, Ltd. Published 2016 by John Wiley & Sons, Ltd.

is just simply to illustrate the nature of the design strategy and provide further direct evidence to support the technique.

Please note that in order to keep this section concise, the maximum levels of induced current as a function of phase has been selected for each frequency and specific plot, and the overall number of plots is restricted in scope.

Figure C.1a–d depicts the induced surface current behaviour for the standard paddles and new designs at 115 MHz for vertical and horizontal polarisations, respectively. The paddles have been modelled in full size – that is, the exact sizes that exist in the real physical chamber.

From Figure C.1a–d it can be clearly seen how the cuts serve to increase the current path length. From the induced surface for the standard paddles it can be seen that at larger wavelengths, the bulk of the surface current is mainly concentrated around the

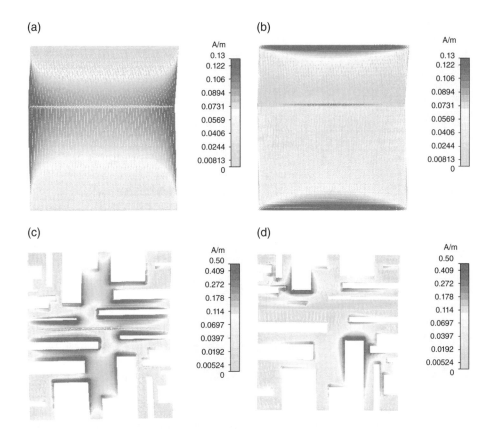

Figure C.1 Induced surface current at 115 MHz: (a) standard paddle vertical polarisation, (b) standard paddle horizontal polarisation, (c) new paddle vertical polarisation and (d) new paddle horizontal polarisation.

outer edges of the paddles in both polarisations. This is because these lengths are the largest dimensions on the paddles and the surface current would appear to be seeking a dimension relative in size in order to interact with the paddles sufficiently.

With the new designs, the overall magnitude of the surface current was a lot higher than for the standard paddles – this is believed to be due to the fact that the new paddle approached a condition of resonance more so than the standard designs. The surface current can be seen to be concentrated around the longer length cuts – that is, these are providing the interaction mechanism between the larger wavelength plane wave and the paddle, which given the overall dimensions of the cuts, would serve to improve the efficiency of the paddles in both polarisations as seen in Chapter 3.

The surface current distributions at 150 MHz can be viewed in Figure C.2a–d.

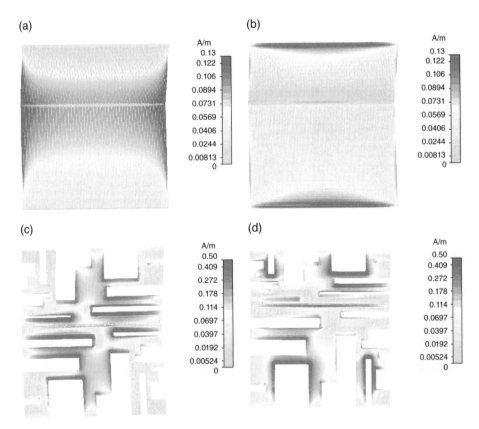

Figure C.2 Induced surface current at 150 MHz: (a) standard paddle vertical polarisation, (b) standard paddle horizontal polarisation, (c) new paddle vertical polarisation and (d) new paddle horizontal polarisation.

From Figure C.2, it can be seen that the current distribution for the standard paddles are largely the same; that is, minimal interaction is expected at 150 MHz similar to 115 MHz. However, when assessing the nature of the surface current for the new designs, in particular for the horizontal polarisation, it is seen that the nature of this induced current is remarkably different. More areas on the paddle are seen to be in use; hence, the multiple cuts of different lengths are serving to provide a resonant capability such that the paddle can sufficiently interact with the plane wave at different wavelengths. This offers evidence in support of the theoretical design strategy from Chapter 3.

Figure C.3a–d illustrates the induced current behaviour at 200 MHz.

The surface current at 400 MHz is shown in Figure C.4.

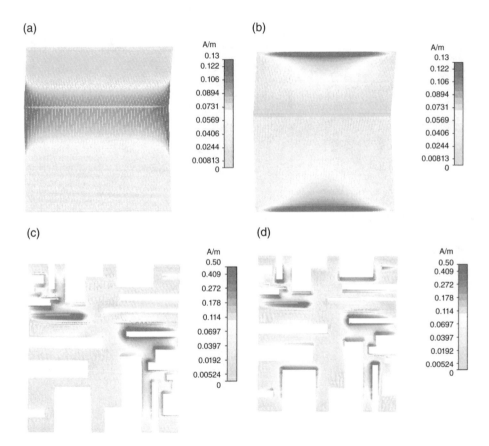

Figure C.3 Induced surface current at 200 MHz: (a) standard paddle vertical polarisation, (b) standard paddle horizontal polarisation, (c) new paddle vertical polarisation and (d) new paddle horizontal polarisation.

(a)   (b)

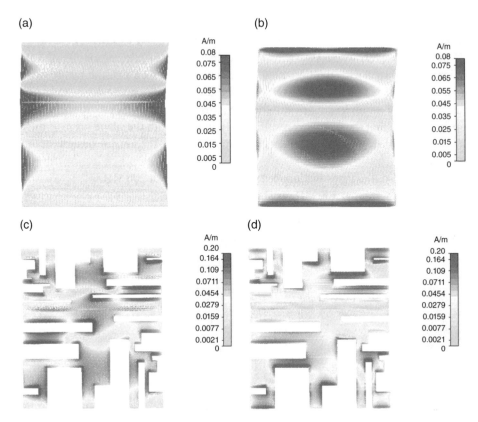

(c)   (d)

Figure C.4 Induced surface current at 400 MHz: (a) standard paddle vertical polarisation, (b) standard paddle horizontal polarisation, (c) new paddle vertical polarisation and (d) new paddle horizontal polarisation.

Appendix D

BS EN 61000-4-21 Standard Deviation Results

As previously stated, this appendix section details the chamber performance advocated by the BS-EN 61000-4-21 standards. The results issued in this section are presented as an illustration for professionals who are familiar with EMC protocols and might be familiar with the results presented in Chapter 3. No weight has been attached to the following results with regards to proving the performance of the new stirrer designs (from Chapter 3) and no significance on the results is intended to be portrayed with respect to the new designs that have been previously presented.

Figures D.1 and D.2 detail the standard deviation for the three (separate) Cartesian field components and the combined (total) electric field as a function of frequency respectively for the standard stirrer paddles (no cuts) in an unloaded chamber. In each figure, the acceptable limits for field uniformity decreed by the BS-EN 61000-4-21 standards are depicted as a solid line. The BS EN 61000-4-21 acceptable limits are that the standard deviation should be within 4 dB below 100 MHz, 4 dB at 100 MHz decreasing linearly to 3 dB at 400 MHz, and within 3 dB above 400 MHz.

From Figures D.1 and D.2, acceptable uniformity is reached at a frequency of 177 MHz.

Moving on to assess the standard paddle design (no cuts) in a loaded chamber, Figures D.3 and D.4 depict the standard deviation for the three (separate) Cartesian field components and the combined electric field as a function of frequency, respectively.

Reverberation Chambers: Theory and Applications to EMC and Antenna Measurements, First Edition.
Stephen J. Boyes and Yi Huang.
© 2016 John Wiley & Sons, Ltd. Published 2016 by John Wiley & Sons, Ltd.

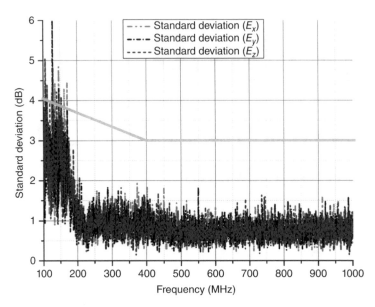

Figure D.1   Standard deviation (dB) for three Cartesian field components from standard stirrer design in an unloaded chamber.

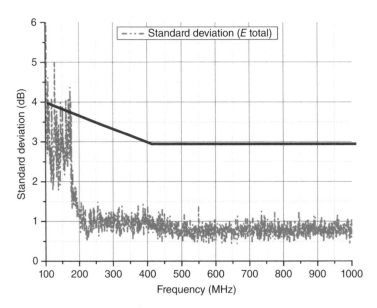

Figure D.2   Standard deviation (dB) for total electric field from standard stirrer design in an unloaded chamber.

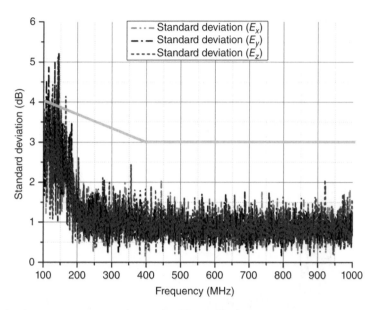

Figure D.3 Standard deviation (dB) for three Cartesian field components from standard stirrer design in a loaded chamber.

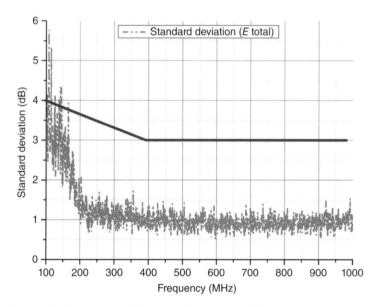

Figure D.4 Standard deviation (dB) for total electric field from standard stirrer design in a loaded chamber.

From Figures D.3 and D.4, acceptable uniformity is reached at a frequency of 170 MHz.

Figures D.5 and D.6 depict the standard deviation performance of the new designs in an unloaded chamber.

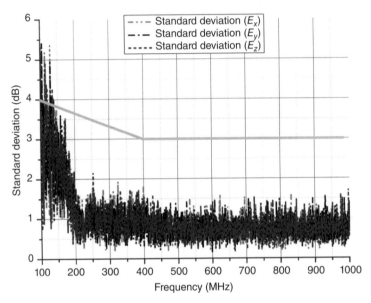

Figure D.5 Standard deviation (dB) for three Cartesian field components from new stirrer design in an unloaded chamber.

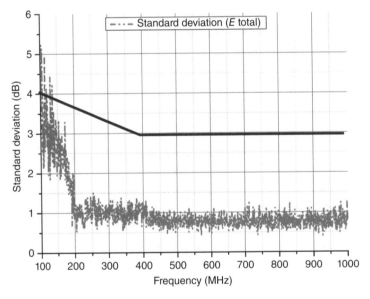

Figure D.6 Standard deviation (dB) for total electric field from new stirrer design in an unloaded chamber.

From Figures D.5 and D.6, acceptable uniformity is reached at a frequency of 139 MHz. The performance of the new designs in a loaded chamber can be viewed in Figures D.7 and D.8.

From Figures D.7 and D.8, acceptable uniformity is reached at a frequency of 136 MHz.

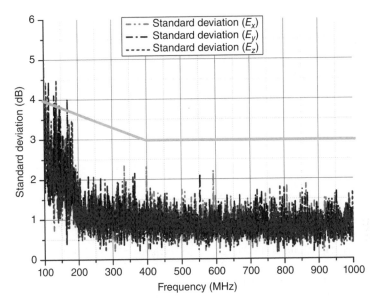

Figure D.7 Standard deviation (dB) for three Cartesian field components from new stirrer design in a loaded chamber.

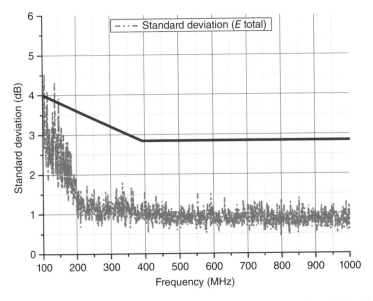

Figure D.8 Standard deviation (dB) for total electric field from new stirrer design in a loaded chamber.

Index

Reverberation Chambers: Theory and Applications to EMC and Antenna Measurements, First Edition.
Stephen J. Boyes and Yi Huang.
© 2016 John Wiley & Sons, Ltd. Published 2016 by John Wiley & Sons, Ltd.